INSECTS AND OTHER TERRESTRIAL ARTHROPODS: BIOLOGY, CHEMISTRY AND BEHAVIOR

BEE HEALTH

BACKGROUND, ISSUES, AND THE ROLE OF PESTICIDES

INSECTS AND OTHER TERRESTRIAL ARTHROPODS: BIOLOGY, CHEMISTRY AND BEHAVIOR

Additional books in this series can be found on Nova's website under the Series tab.

Additional e-books in this series can be found on Nova's website under the e-book tab.

INSECTS AND OTHER TERRESTRIAL ARTHROPODS: BIOLOGY, CHEMISTRY AND BEHAVIOR

BEE HEALTH

BACKGROUND, ISSUES, AND THE ROLE OF PESTICIDES

CRISTINA WEAVER
EDITOR

nova publishers
New York

NOTICE TO THE READER

Library of Congress Cataloging-in-Publication Data

ISBN: 978-1-63482-671-6

Published by Nova Science Publishers, Inc. † New York

CONTENTS

Preface		**vii**
Chapter 1	Bee Health: Background and Issues for Congress *Renée Johnson and M. Lynne Corn*	**1**
Chapter 2	Bee Health: The Role of Pesticides *Renée Johnson and M. Lynne Corn*	**47**
Chapter 3	Guidance for Assessing Pesticide Risks to Bees *Office of Pesticide Programs, EPA*	**107**
Chapter 4	Statement of Jeff Pettis, Research Leader, USDA Agricultural Research Service. Hearing "To Review Current Research and Application of Management Strategies to Control Pests and Diseases of Pollinators"	**181**
Chapter 5	Testimony of Arthur D. Cummings, CEO, Capay Farms. Hearing "To Review Current Research and Application of Management Strategies to Control Pests and Diseases of Pollinators"	**187**
Chapter 6	Testimony of Jeff Stone, Executive Director, Oregon Association of Nurseries. Hearing "To Review Current Research and Application of Management Strategies to Control Pests and Diseases of Pollinators"	**193**
Index		**201**

PREFACE

This book provides a background on bee health. It also discusses the role of pesticides in bee health and provides guidance to risk assessors for evaluating the potential risk of pesticides to bees, particularly honey bees.

Chapter 1 – Bees, both commercially managed honey bees and wild bees, play an important role in global food production. In the United States alone, the value of insect pollination to U.S. agricultural production is estimated at $16 billion annually, of which about three-fourths is attributable to honey bees. Worldwide, the contribution of bees and other insects to global crop production for human food is valued at about $190 billion. Given the importance of honey bees and other bee species to food production, many have expressed concern about whether a "pollinator crisis" has been occurring in recent decades.

Over the past few decades there has been heightened concern about the plight of honey bees as well as other bee and pollinator species. Although honey bee colony losses due to bee pests, parasites, and disease are not uncommon, there is the perception that bee health has been declining at a faster rate both in the United States and globally in recent years. This situation gained increased attention in 2006 as some commercial beekeepers began reporting sharp declines in their honey bee colonies. Because of the severity and unusual circumstances of these colony declines, scientists named this phenomenon colony collapse disorder (CCD). Since then, honey bee colonies have continued to dwindle each year, for reasons not solely attributable to CCD. The U.S. Department of Agriculture (USDA) reports that CCD may not be the only or even the major cause of bee colony losses in recent years. In the United States, USDA estimates of overwinter colony losses from all causes have averaged more than 30% annually since 2006.

To date, the precise reasons for bee colony losses are not yet known. Reasons cited for bee declines include a wide range of possible factors thought to be affecting pollinator species. These include bee pests and disease, diet and nutrition, genetics, habitat loss and other environmental stressors, agricultural pesticides, and beekeeping management issues, as well as the possibility that bees are being affected by cumulative, multiple exposures and/or the interactive effects of several of these factors.

Chapter 2 – Reports worldwide indicate that populations of managed honey bees, wild bees, and native bees have been declining, with colony losses in some cases described as severe or unusual. Other reports indicate that many insect pollinator species may be becoming rarer, which some say may be a sign of an overall global biodiversity decline.

Many reasons are cited for bee population declines, including bee pests and diseases, diet and nutrition, genetics, habitat loss, agricultural pesticides, and beekeeping management. Because pesticides have been the focus of concerns in Europe and in the United States, this CRS report briefly describes recent scientific research and analysis regarding the potential role of pesticides among the factors affecting the health and well-being of bees. The report concludes with a summary of recent regulatory activity regarding neonicotinoids, a type of pesticide, and also provides the statutory authority and regulatory activities related to pesticide use at the U.S. Environmental Protection Agency (EPA), the federal agency charged with assessing risks and regulating U.S. sale and use of pesticides.

Chapter 3 – The EPA's Pollinator Risk Assessment Guidance is part of a long-term strategy to advance the science of assessing the risks posed by pesticides to bees, giving risk managers the means to further improve pollinator protection in regulatory decisions. Among other things, the EPA anticipates the guidance will allow the agency to assess effects from systemic pesticides quantitatively on individual bees as well as on bee colonies. This guidance is an outgrowth of the 2012 FIFRA Scientific Advisory Panel Meeting on Risk Assessment for Bees.

Chapter 4 – This is the Statement of Jeff Pettis, Research Leader, USDA Agricultural Research Service. Hearing "To Review Current Research and Application of Management Strategies to Control Pests and Diseases of Pollinators", date April 29, 2014.

Chapter 5 – This is the Testimony of Arthur D. Cummings, CEO, Capay Farms. Hearing "To Review Current Research and Application of Management Strategies to Control Pests and Diseases of Pollinators", date April 29, 2014.

Chapter 6 – This is the Testimony of Jeff Stone, Executive Director, Oregon Association of Nurseries. Hearing "To Review Current Research and Application of Management Strategies to Control Pests and Diseases of Pollinators", date April 29, 2014.

In: Bee Health
Editor: Cristina Weaver

ISBN: 978-1-63482-671-6

Chapter 1

BEE HEALTH: BACKGROUND AND ISSUES FOR CONGRESS*

Renée Johnson and M. Lynne Corn

SUMMARY

Bees, both commercially managed honey bees and wild bees, play an important role in global food production. In the United States alone, the value of insect pollination to U.S. agricultural production is estimated at $16 billion annually, of which about three-fourths is attributable to honey bees. Worldwide, the contribution of bees and other insects to global crop production for human food is valued at about $190 billion. Given the importance of honey bees and other bee species to food production, many have expressed concern about whether a "pollinator crisis" has been occurring in recent decades.

Over the past few decades there has been heightened concern about the plight of honey bees as well as other bee and pollinator species. Although honey bee colony losses due to bee pests, parasites, and disease are not uncommon, there is the perception that bee health has been declining at a faster rate both in the United States and globally in recent years. This situation gained increased attention in 2006 as some commercial beekeepers began reporting sharp declines in their honey bee colonies. Because of the severity and unusual circumstances of these

* This is an edited, reformatted and augmented version of a Congressional Research Service publication R43191, prepared for Members and Committees of Congress, dated January 20, 2015.

colony declines, scientists named this phenomenon colony collapse disorder (CCD). Since then, honey bee colonies have continued to dwindle each year, for reasons not solely attributable to CCD. The U.S. Department of Agriculture (USDA) reports that CCD may not be the only or even the major cause of bee colony losses in recent years. In the United States, USDA estimates of overwinter colony losses from all causes have averaged more than 30% annually since 2006.

To date, the precise reasons for bee colony losses are not yet known. Reasons cited for bee declines include a wide range of possible factors thought to be affecting pollinator species. These include bee pests and disease, diet and nutrition, genetics, habitat loss and other environmental stressors, agricultural pesticides, and beekeeping management issues, as well as the possibility that bees are being affected by cumulative, multiple exposures and/or the interactive effects of several of these factors.

USDA continues to research possible causes of bee colony losses, and has published a series of reports detailing the agency's progress in this area. In 2013, USDA and the U.S. Environmental Protection Agency (EPA) published a joint report, *National Stakeholders Conference on Honey Bee Health*. Both USDA and the NAS report conclude that many factors contribute to pollinator declines in North America. A 2007 report by the National Research Council, *Status of Pollinators in North America,* also provides a detailed scientific context concerning bee health.

Following heightened concern over honey bee colony losses in 2006-2007, Congress provided for increased funding for bee research, among other types of farm program support to protect pollinators, as part of the 2008 farm bill (P.L. 110-246). The 2014 farm bill (P.L. 113-79) reauthorized and expanded many of these provisions, addressing managed honey bees and native pollinators as part of the law's research, conservation, specialty crop, and miscellaneous title provisions. In addition, outside the farm bill, H.R. 4790 would promote conservation practices on millions of acres of highway rights-of-way by encouraging states to reduce mowing and plant for pollinators, providing improved habitat for pollinators and other small wildlife. Also, H.R. 2692 would suspend registrations of neonicotinoids and prohibit new registrations of any pesticide for use unless EPA determines the insecticide would not cause unreasonable adverse effects on pollinators, including honey bees and native bees as well as other pollinators. Another bill, H.R. 5447, would amend U.S. pesticide laws to provide for expedited registration of pesticides that improve bee health, including managing resistance to some parasitic pests.

Bees, both commercially managed honey bees and wild unmanaged bees, among many other types of insect pollinators, play an important role in global food production. In the United States, the value of pollination by bees and other insects to U.S. agricultural production is estimated at $16 billion annually.[1] Most of this estimated value (about 75%) is attributable to honey bees.

Given the importance of bees and other types of pollinators to food production, many have expressed concern about whether a "pollinator crisis" has been occurring in recent decades. Worldwide reports indicate that populations of both managed honey bees and native bees have been declining, with colony losses in some cases described as severe or unusual. In Europe, managed honey bee colony numbers have been declining since the mid-1960s, and individual beekeepers have reported "unusual weakening and mortality in colonies," particularly during the period spanning winter through spring.[2] According to the United Nations, many insect pollinator species may be becoming rarer, causing some to question whether this is a sign of an overall global biodiversity decline.[3]

In the United States, honey bee colony losses due to bee pests, parasites, pathogens, and disease are not uncommon. This situation gained increased attention in 2006 as some commercial beekeepers began reporting sharp declines in their honey bee colonies. Because of the severity and unusual circumstances of these colony declines, scientists named this phenomenon colony collapse disorder (CCD). This issue was legislatively active in the 110th Congress and resulted in increased funding for pollinator research, among other types of farm program support to protect pollinators, as part of the 2008 farm bill (P.L. 110-246).[4]

Years later, the U.S. Department of Agriculture (USDA) reports that U.S. beekeepers continue to lose honey bee colonies each year. Since 2006, USDA estimates that overwinter colony losses have averaged more than 30% annually.[5] However, USDA reports that bee colonies have continued to decline each winter for reasons not solely attributable to CCD, which is characterized by colony populations that are suddenly lost but no dead bees are present. (For more information on CCD, see text box on next page. Additional discussion is provided in "Factors Affecting Bee Health.") USDA claims that "beekeepers did not report CCD as a major cause of colony loss" for overwinter losses in most recent surveys.[6] To date, the precise reasons for honey bee colony losses are still unknown.

Colony Collapse Disorder (CCD)

In late 2006, commercial beekeepers along the East Coast began reporting sharp declines in their honey bee colonies. Because of the severity and unusual circumstances of these colony declines, scientists named this phenomenon colony collapse disorder (CCD).

According to USDA: "The defining characteristic of CCD is the disappearance of most, if not all, of the adult honey bees in a colony, leaving behind honey and brood (immature bees confined to cells in the hive, including larvae and pupae) and the absence or dead bees (i.e., no dead bee bodies are present)." Symptoms include the following:

- rapid loss of adult worker bees,
- few or no dead bees found in the hive,
- presence of immature bees (brood),
- small cluster of bees with live queen present, and
- pollen and honey stores in hive.

Another defining characteristic that may be indicative of CCD is low levels of *Varroa* mite and other pathogens, such as *Nosema*, which are generally associated with most winter colony losses.

Since the honey bee is a social insect and colony-oriented, with a complex and organized nesting colony, failing to return to the hive is considered highly unusual. Also, there is little evidence that the hive may have been attacked. In actively collapsing colonies, an insufficient number of adult bees remain to care for the brood. An absence of a large number of dead bees makes an analysis of the causes of CCD difficult. To date, the precise reasons for honey bee colony losses attributable to CCD are still unknown.

Reasons cited for bee colony declines include a wide range of possible factors thought to be negatively affecting pollinator species: bee pests and disease, diet and nutrition, genetics, habitat loss and other environmental stressors, agricultural pesticides, and beekeeping management issues, as well as the possibility that bees are being negatively affected by cumulative, multiple exposures and/or the interactive effects of each of these factors.

As reported by USDA: "While many possible causes for CCD have been proposed, reported, and discussed—both in the scientific literature and popular media—no cause has been proven."

Scientists at USDA and bee labs across the country continue to look for the cause or causes of CCD within four broad categories: (1) pathogens (such as *Nosema ceranae*); (2) parasites (such as *Varroa* mites); (3) environmental stressors (such as pesticides or lack of nectar diversity); and (4) management stressors (such as transportation stress by migratory beekeepers).

Sources: K. Kaplan, "Colony Collapse Disorder: An Incomplete Puzzle," *Agricultural Research* (USDA publication), July 2012; K. Kaplan, "Fact Sheet: Survey of Bee Losses during Winter of 2012/2013," May 2013; D. Evans, Yan Ping Chen, et al., "Predictive markers of honey bee colony collapse," *PLoS ONE*, vol. 7, no. 2 (p. e32151), 2012; NRC, *Status of Pollinators in North America*, 2007; USDA, *Colony Collapse Disorder Progress Report*, various years; and comments by Laurie Adams, North American Pollinator Protection Campaign (NAPPC), in a presentation to congressional staff on June 18, 2012. For other information, see CRS Report RL33938, *Honey Bee Colony Collapse Disorder*.

Reasons cited for honey bee population losses include a wide range of possible factors. Potential identified causes include bee pests and disease, diet and nutrition, genetics, habitat loss and other environmental stressors, agricultural pesticides, and beekeeping management issues, as well as the possibility that bees are being negatively affected by cumulative, multiple exposures and/or the interactive effects of several of these factors.

Given continued concerns about the health and well-being of honey bees and other pollinators, this issue has continued to be legislatively active in Congress. Various proposals addressing managed bees and native pollinators were enacted in the 2008 farm bill (P.L. 110-246) and reauthorized and expanded as part of the 2014 farm bill (P.L. 113-79), in the law's research, conservation, specialty crop, and miscellaneous title provisions.[7]

OVERVIEW OF SELECTED BEE SPECIES

More than 200,000 species of insects (bees, butterflies, moths, flies, wasps, beetles, and ants), bats (and other small mammals), and birds (including hummingbirds) are plant pollinators. This report focuses on bees—both managed honey bees and wild (mostly unmanaged) bee species.

Managed Honey Bees

Honey bees (*Apis mellifera*; Family: Apidae) are the most well-known bee species. However, honey bees are only one of the world's estimated 17,000 described bee species, and one of the estimated total of 20,000 to 30,000 bee species worldwide.[8] Honey bees are not native to North America, but were introduced by European settlers in the 1600s.

Honey bees are considered to be "social" bees in that they have single egg-laying queen and sterile worker bees that tend to work together in a highly structured social order, consisting of cooperation and division of labor within a colony, as well as the presence of two generations in a single nest at the same time. Social behavior allows bees to be domesticated and managed.

There are an estimated 115,000-125,000 beekeepers in the United States.[9] Most of these (roughly 90,000-100,000) are hobbyists with fewer than 25 hives.[10] Commercial beekeepers tend to have more than 300 hives. They migrate their colonies during the year to provide pollination services to farmers. In the United States, most pollination services are provided by commercial beekeepers.[11]

Some other types of native bees are also managed, such as bumble bees, orchard bees, and alfalfa leaf-cutting bees. Some of these bees (such as bumble bees and some types of stingless bees) exhibit some of the social behaviors commonly associated with honey bees. Some bumble bee species are managed in controlled environments to pollinate greenhouse tomatoes. Others are valued as primary or supplementary pollinators for members of the squash family such as cucumbers, watermelons, and cantaloupes. Some species of meliponine bees (commonly called stingless bees), which are mostly found in the tropics, were widely cultivated by Native Americans in Central and South America before the introduction of honey bees from Europe.[12] Roughly 500 species of bees native to North America show some degree of social behavior.

Wild Bee Species

An estimated 4,000 species of bees are native to North America.[13] With few exceptions, most of these are wild and not managed. Most types of wild bees are "solitary" bees—estimated at about 3,500 species—and do not have long-lived colonies.[14] Data and information on native bees are more limited than information on managed bees.

The five most common families of native bees in North America are Andrenidae, Apidae, Colletidae, Halictidae, and Megachilidae. Andrenid bees are all ground nesters, and mostly comprise a large family of dark, nondescript bees, although some are colorful. Bees in the large Apidae family include not only honey bees, but also bumble bees (such as *Bombus* spp.), carpenter bees, squash or gourd bees, and others. Compared to most native species, bumble bees (*Bombus)* are better studied. There are 239 species of bumble bees worldwide; of these, 60 species are found in the United States, Mexico, and Canada.[15]

Classification of Bees

Kingdom Animalia—Animal
- Phylum Arthropoda—Arthropods
  - Class Insecta—insects, hexapoda
    - Order Hymenoptera—ants, bees, wasps
      - Superfamily Apoidea—bees, sphecoid wasps, apoid wasps
        - *Family* Andrenidae—andrenid bees, andrenids
        - *Family* Apidae—honey bees, bumble bees, carpenter bees, squash bees, stingless bees
        - *Family* Colletidae—colletid bees, plasterer bees, yellow-faced bees
        - *Family* Halictidae—halictid bees, sweat bees
        - *Family* Megachilidae—leafcutting bees, orchard bees, mason bees, resin bees
        - *Family* Melittidae—melittid bees, melittids

Source: Integrated Taxonomic Information System (ITIS). ITIS is supported by a consortium of federal and international agencies and scientific organizations to provide authoritative taxonomic information on known plant and animal species. Available at http://www.itis.gov/. Condensed and slightly modified by CRS to include more common names.

Bees in the smaller Colletidae family are broader and wasp-like, and include plasterer bees. Bees in the Halictidae family include many species of sweat bees, and most have shiny metallic colored or black bodies. Bees in the Megachilidae family include resin and mason bees, orchard bees, and leaf-cutter bees that mostly nest in holes. Most families of bees contain some types of parasitic and cuckoo[16] bees. For a summary of the different types of wild bees see the text box above.

Importance of Bee Pollination

Worldwide, the contribution of bees and other insects to worldwide crop production for human food is valued at about $190 billion each year.[17] In the United States alone, the value of insect pollination to U.S. agricultural production is estimated at $16 billion annually, of which about three-fourths of the value is attributable to honey bees.[18]

Expressed somewhat differently, another widely cited estimate is that bee pollination of agricultural crops is said to account for about one-third of the U.S. diet, contributing to the production of a diverse range of high-value fruits, vegetables, tree nuts, forage crops, some field crops, and other specialty crops.[19]

Both managed and wild bees are critical to plant pollination and are economically valuable to U.S. agricultural production. Bee pollination of agricultural crops is said to contribute to the production of as many as 90 agricultural crops.[20] A number of agricultural crops are almost totally (90%-100%) dependent on animal pollination, including apples, avocados, blueberries, cranberries, cherries, kiwi fruit, macadamia nuts, asparagus, broccoli, carrots, cauliflower, celery, cucumbers, onions, legume seeds, pumpkins, squash, and sunflowers.[21] Other specialty crops also rely on pollination, but to a lesser degree.[22] USDA reports that native bees also provide pollination services for a number of food crops. In addition to some of the aforementioned crops, these include alfalfa seeds, almonds, canola, chokecherries, grapefruit, pears, plums, prunes, soybeans (hybrid seed production), tomatoes, vegetable seeds, and watermelons.[23] The role of bee pollination in enhancing agricultural productivity and crop yields for many types of fruit, vegetable, tree nut, and field crops was also noted in a series of published studies as part of the United Nations' Global Pollination Project.[24]

Native bees are thought to be more efficient at pollination than honey bees and are known to pollinate native plants, such as cherries, blueberries, and cranberries, as well as other commercial crops including almonds, among other fruiting and flowering plants.[25]

Estimated Economic Value

The economic value of pollination services provided by managed honey bees and wild bees is difficult to quantify. Most available estimates focus on crop pollination by managed honey bees. Data and information on native bees are more limited. Where few bee species are kept commercially and managed for their pollination services, wild bee species also play an important role in providing pollinator services to both commercial and small-scale home gardening systems. Both locally and globally, some are concerned that the availability of honey bee stocks is not keeping pace with growing agricultural demands for pollination services.[26]

Researchers at Cornell University estimate the monetary value to U.S. crops of total insect pollination—both commercial honey bees and other insect pollinators such as non-*Apis* bees and other insects—at about $16.4 billion in 2010.[27] (*Table 1*) Of the estimated $16.4 billion attributable to insect pollination in the United States, the value of pollination from honey bees is estimated at $12.4 billion, or about three-fourths of the total value. The value of pollination from other insects, including non-*Apis* bees and other insects, is estimated at $4.0 billion.[28]

Separate estimates of the economic value of only wild bee pollination services are not readily available. A 2011 study by researchers at the University of California estimated that wild bee species add $0.9 billion to $2.4 billion per year in value to California's agriculture through pollination services.[29] Some researchers estimate that native bees contribute up to 15% of the value of all pollination services to agricultural and native landscapes.[30] Additional efforts are underway in an attempt to estimate the value of pollination services of wild bees.

Commercial Migratory Beekeeping

In the United States, most commercial pollination services are provided by migratory beekeepers that travel from state to state and provide pollination services to crop producers. These operations are able to supply a large number of bee colonies during the critical phase of a crop's bloom cycle, when honey bees pollinate a crop as they fly from flower to flower collecting nectar and pollen, which they carry back to the nest.[31] Of the roughly 2 million bee colonies rented out each year to pollinate U.S. agricultural crops, an estimated 1.6 million colonies are needed to pollinate California's more than 900,000 acres of almond trees alone.[32] Increasingly, bee colony rentals are also being used to pollinate cultivated blueberries in the East Coast states and in the Midwest.[33] Other information indicates that bee colonies are also rented for apple, pear, plum, cherry, cranberry, avocado, cucumber, kiwi fruit, melon, pepper, and citrus fruit production, as well as for alfalfa, clover seed, and sunflower production. (See map in *Figure 1.)* Overall, demand for pollination services is rising faster than available supplies, which are further exacerbated by declining managed and wild pollinator populations.[34]

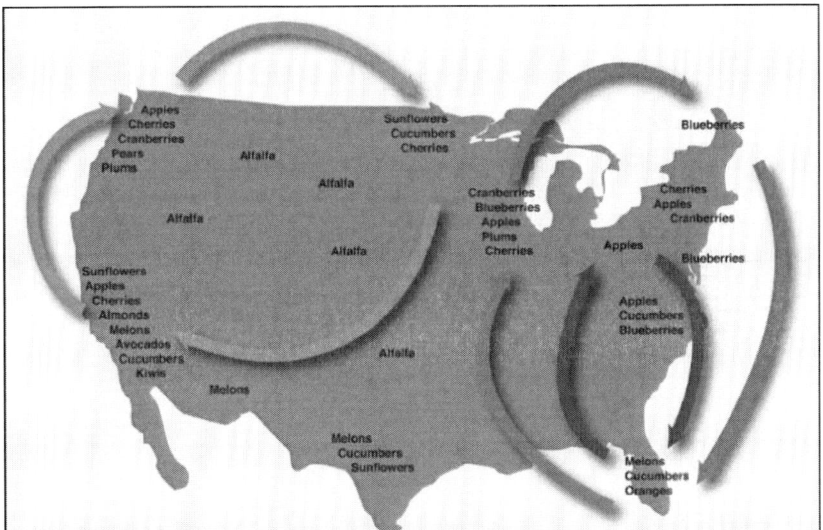

Source: Posted at Person County (NC) Beekeepers Association by Inge Kautzmann, October 28, 2011. See also Penn State University, "A Year in the Life of a Migratory Honey Bee Colony"; M. Burgett, 1999 Pacific Northwest Honey Bee Pollination Survey, Oregon State University; and Pollination Map for Honey Bees in the United States.

Figure 1. Map of Commercial Pollination.

Rental fees collected by commercial beekeepers for pollination services may vary by crop type, and tend to be lower for some seed crops and higher for berry and tree crops. In recent years, pollination fees paid by crop producers have increased. For example, fees paid by California's almond industry have risen from a reported $35 per colony in the late 1990s to about $75 per colony in 2005. More recent estimates of fees for pollinating almond trees are even higher, at $150 per colony or more (*Figure 2*).[35] Fees may also vary by region.

Table 1. Estimated Value of Insect Pollination to U.S. Crop Production, 2010

Crop Category	Dependence on Insect Pollination	Proportion of Pollinators That Are Honey bees	Value Attributed to All Insect Pollinators ($ millions)	Portion Attributed to Honey bees ($ millions)	Major Producing States
Grain and seed crops	10%-100%	20%-90%	826.9	744.2	CA, ND, SD, ID, WI
Peanuts	10%	50%	90.1	18.0	GA, FL, AL, TX, NC
Cotton (seed)	20%	80%	200.8	160.6	TX, AR, GA, MS
Apples	100%	90%	2,220.8	1,998.7	WA, NY, MI, PA
Peaches, nectarines	60%	80%	446.4	357.1	CA, SC, GA, PA, WA
Prunes, plums	70%	90%	163.2	146.9	CA, ID, MI, OR, WA
Cherries	90%	90%	685.5	617.0	WA, CA, OR, MI
Other fruit (incl. olives)	10%-90%	10%-90%	656.3	578.3	CA, FL, WA
Almonds	100%	100%	2,838.5	2,838.5	CA
Macadamia nuts	90%	90%	27.0	24.3	HI
Blueberries	100%	90%	644.0	579.6	ME, MI, GA, OR, WA, NJ
Strawberries	20%	10%	449.1	44.9	CA, FL, OR
Other berries	80%-100%	90%	551.7	496.5	WA, CA, OR
Oranges	30%	90%	580.5	522.4	CA, FL, AZ, TX
Grapefruit	80%	90%	228.8	205.9	FL, CA, AZ, TX
Lemons	20%	10%	76.1	7.6	CA, AZ
Tangerines, tangelos	40%-50%	90%	140.8	126.7	CA, FL
Soybeans	10%	50%	3,891.5	1,945.8	IA, IL, MN, IN

Table 1. (Continued)

Crop Category	Dependence on Insect Pollination	Proportion of Pollinators That Are Honey bees	Value Attributed to All Insect Pollinators ($ millions)	Portion Attributed to Honey bees ($ millions)	Major Producing States
Cantaloupes	80%	90%	251.5	226.4	CA, AZ, GA, CO, IN, PA
Cucumbers	90%	90%	340.4	306.3	FL, GA, NC, CA, NY, MI
Squash	90%	10%	183.2	18.3	FL, NY, CA, NC
Other melons, pumpkins	70%-90%	10%-90%	489.0	356.2	IL, CA, NY, OH, PA, MI
Grapes	10%	10%	362.7	36.3	CA, OR, WA, VA
TOTAL	—	—	$16,345	$12,357	—

Source: Compiled by CRS using values reported in N. W. Calderone, "Insect Pollinated Crops, Insect Pollinators and US Agriculture: Trend Analysis of Aggregate Data for the Period 1992–2009," PLoS ONE 7(5): e37235, May 2012, http://www.plosone.org/article/info%3Adoi%2F10.1371%2Fjournal.pone.0037235. Supporting documentation provides data for individual crops from 1992 through 2009, with an update for 2010. Updates previous estimates by R. A. Morse and N. W. Calderone, The Value of Honey Bees as Pollinators of U.S. Crops in 2000, Cornell University, March 2000.

Notes: Value of crop production attributable to insect pollinators, including honey bees, is expressed in terms of the plant's dependence on pollination, and the insect's contribution to increased yield and quality achieved from pollination, including the indirect benefits of insect pollination required for seed production of some crops.

Among the reasons for higher pollination fees are expanding almond acreage and relatively high honey prices, but also fewer available honey bees for pollination due, in part, to colony declines and bee mortalities. In 2009, across a range of commodities (excluding almonds), average rental fees for pollination services ranged from $39 per colony to $70 per colony.[36]

Precise statistics are not available on the number of commercial migratory beekeepers. In the United States, it is generally believed that operations with the majority of bee colonies—more than 2 million of the roughly 3 million bee colonies in the United States—belong to commercial migratory beekeepers, mostly in the Dakotas and in some midwestern states.

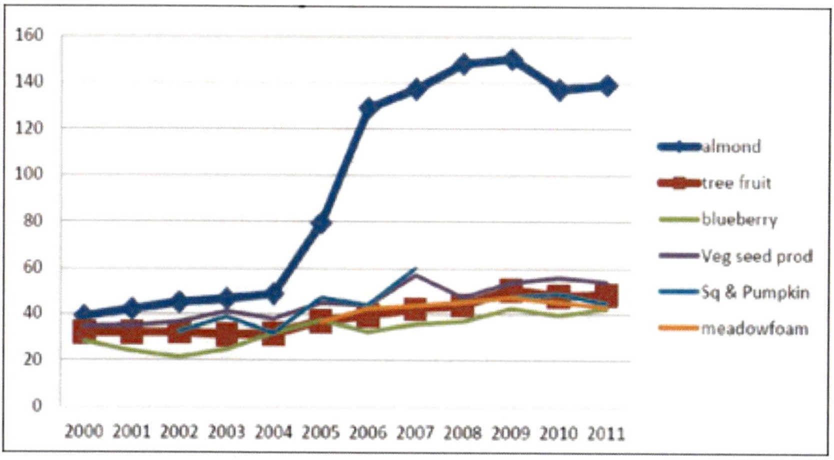

Source: D. M. Caron, R. Sagili, and M. Cooper, "Pacific Northwest (PNW) 2011 Beekeeper Pollination."

Notes: "Weighted fee" = total rental value divided by number of rental colonies. Data for Eastern U.S. states are at D. M. Caron, "Bee Colony Pollination Rental Prices, Eastern U.S. with Comparison to West Coast," 2011.

Figure 2. Weighted Pollination Fee, by Major Sector, Northwest, 2000-2011. (total rental value divided by number of rental colonies [$/colony])

Changes in Bee Populations

Managed Honey Bees

USDA does not compile comprehensive annual survey data on honey bee colonies. Available data are limited and not ideal for evaluating population changes among honey bees. Annual data on the number of honey bee colonies are from statistics tracking the number of operations that produce honey for commercial sale. Additional data on all "colonies of bees" are from USDA's five-year *Census of Agriculture* (the most recent available data are for 2012). In response to concerns about honey bee declines, USDA and the national bee laboratories have started to survey beekeepers each year regarding overwinter colony losses as part of the "Bee Informed Partnership."

Available USDA data on the number of bee colonies in honey production are not strictly comparable. Data are compiled for different purposes—the *Census* data track all farms with bee colonies, the annual honey production data track operations that produce honey for commercial sale. Data are also compiled using different data collection techniques—one is a periodic census, the other a statistical estimate. Generally, USDA data are mostly intended to track the number of honey-producing colonies; however, it is the value of managed bees as crop pollinators that provides the greatest economic impact in the production of food and feed crops.

A 2007 study by the National Research Council (NRC) voiced concerns about the available USDA data on honey bees, cautioning against use of these data to determine changes in honey bee populations.[37] Concerns about the data include the following: the data track U.S. honey production rather than bee populations; the lack of comparability among the available data compilations; the potential for misidentification of species and miscounting because of data collection procedures; and other issues.

Notwithstanding these concerns about data availability and quality, the NRC study concluded: "Long-term population trends for the honey bee, the most important managed pollinator, are demonstrably downward."[38] Honey bee colony losses are not uncommon. The NRC study cites USDA data showing honey bee declines in 1947–1972 and 1989–1996, as well as declines starting in 2005 (despite reports of a sharp rise in 2010).

U.S. Farms with Bee Colonies

Data from the most recent USDA *Census of Agriculture* indicate that there were 3.3 million bee colonies on U.S. farms in 2012.[39] These data are based on the reported 38,300 farms with honey production during that year. This compares to previous *Census* data from 2007, when there were 28,000 operations with 2.9 million bee colonies in the United States, reflecting the recent increased interest in beekeeping. Available limited historical data indicate that there were an estimated 6 million honey bee colonies on U.S. farms during the 1950s.[40] These overall trends may reflect ongoing consolidation and structural shifts in the U.S. agricultural sectors, rather than conclusive trends in species populations, abundance, and distribution.

By state, about one-third of all bee colonies are located in California, with another roughly one-third located in in North and South Dakota, Florida, Texas, and Montana. Other states include Washington, Oregon, Michigan, New York, Georgia, Arizona, Wisconsin, Nebraska, and Wyoming. Although these operations also produce honey for commercial sale, their value as crop pollinators provides the greatest economic impact in the production of food and feed crops.

Honey-Producing Bee Colonies

USDA annual data for honey-producing colonies are similarly inconclusive. These data show year-to-year fluctuations, but not a clear downward trend over time.

Over the 10-year period shown, the number of honey-producing bee colonies dropped from 2.6 million bee colonies in 2002-2003 to lows of 2.3 million to 2.4 million bee colonies in 2006- 2008. In 2010, the sector reached a 10-year high when honey production increased and the number of honey bee colonies totaled nearly 2.7 million (*Table 2*).

Table 2. Honey-Producing Bee Colonies, 2002-2013

Year	Number of Honey-Producing Colonies (1,000 colonies)	Honey Production (1,000 pounds)
2002	2,574	171,718
2003	2,599	181,727
2004	2,556	183,582
2005	2,413	174,818
2006	2,393	154,907
2007	2,443	148,341
2008	2,342	163,789
2009	2,498	146,416
2010	2,692	176,462
2011	2,491	148,357
2012	2,539	142,296
2013	2,640	149,499

Source: USDA, Honey, http://usda.mannlib.cornell.edu/MannUsda/view Document Info.do?documentID=1191. Honey production from producers with five or more colonies. Honey producing colonies are the maximum number of colonies from which honey was taken during the year, although it is possible to take honey from colonies that did not survive the entire year.

This increase in the number of honey-producing hives may reflect efforts among beekeepers to continually replenish their hives—either in response to a colony die-off or to increase the number of hives to raise overall honey production or to provide additional pollinator services. This increase might also reflect the rising popularity of beekeeping as a part-time hobby.

Overall, these data illustrate that despite a greater number of honey-producing colonies, annual honey production remains lower compared to just a decade ago.

Measured by overall honey production in 2013, states producing the most honey in the United States include North Dakota (22%), South Dakota (10%), Montana (10%), Florida (9%), California (7%), and Minnesota (5%).[41] Other major honey-producing states are Idaho, Texas, Michigan, Oregon, Georgia, Nebraska, New York, Washington, Wisconsin, and Wyoming.

Annual "Winter Loss" Survey

Heightened attention about CCD has resulted in better tracking of annual honey bee colony losses by USDA and national bee laboratories. This effort has evolved into the collection of an annual winter loss survey, conducted by the Bee Informed Partnership[42] and the Apiary Inspectors of America (AIA), along with leading research labs and universities in agriculture and science, and is supported by USDA and its research agency, the National Institute of Food and Agriculture (NIFA).[43]

USDA reports that bee colony losses averaged 17%-20% per year between the 1990s and mid-2000s, attributable to a variety of factors, such as mites, diseases, and management stress.[44] By comparison, bee colony losses between the winters of 2006/2007 and 2013/2014 have averaged nearly 30% annually (*Figure 3*). Reported annual loss rates of managed honey bee colonies from all causes nationwide are as follows:[45]

- 23% in the winter of 2013/2014;
- 31% in the winter of 2012/2013;
- 22% in the winter of 2011/2012;
- 30% in the winter of 2010/2011;
- 34% in the winter of 2009/2010;
- 29% in the winter of 2008/2009;
- 36% in the winter of 2007/2008;
- 32% in the winter of 2006/2007.

Other information from USDA's 2012/2013 annual survey indicates that 70% of responding beekeepers reported losses greater than 14%—the level of loss that beekeepers say allow them to remain economically viable as a business.[46]

The first survey in this series was conducted by the AIA, covering the winter of 2006/2007.[47] Of the responding beekeepers, about one-fourth reported conditions associated with CCD.[48] The survey indicated that, among the beekeepers surveyed, more than 50% reported "abnormally heavy losses" with total colony losses of 55%. This compared to those reporting "normal losses" with total colony losses of 16%. Beekeeping operations experiencing CCD-like conditions reported losses of 45% of their managed bee colonies. Among the leading causes reported by most affected commercial beekeeping operations were pest diseases.

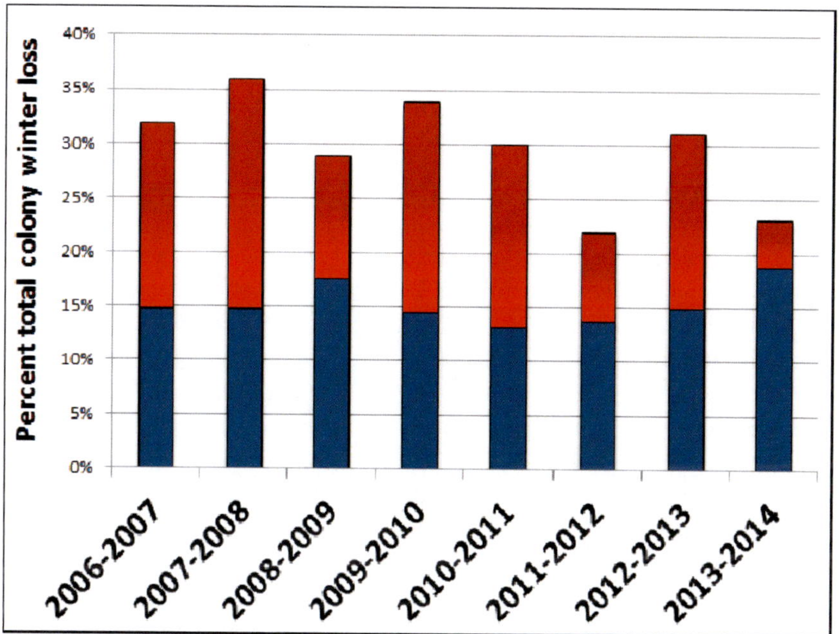

Source: D. vanEngelsdorp et al., "Colony Loss 2013-2014," May 15, 2014. Summary of total overwinter colony loss (October 1 – April 1) of managed honey bee colonies in the United States across the 8 annual national surveys (red bars). The acceptable range (blue bars) is the average percentage of acceptable loss declared by the survey participants in each of the eight years of the survey.

Figure 3. Managed Honey Bee Colony Losses in the United States.

Subsequent survey results indicate that the number of managed honey bee colonies again dropped in each of the following winters. One exception was overwinter 2011/2012, when surveyed total losses of colonies from all causes were 22% nationwide and represented a substantial improvement in mortality compared to the previous five years. The following year (2012/2013), however, total losses of managed honey bee colonies nationwide again reached more than 30% from all causes. Estimated losses in overwinter 2013/2014 dropped back again to 23%.

Based on the annual winter loss surveys, USDA estimates that overwinter colony losses have averaged more than 30% annually since 2006/2007. However, bee colonies have continued to "dwindle away" each year for reasons not solely attributable to CCD, which is characterized by colony populations that are suddenly lost.[49] USDA claims that "beekeepers did not report CCD as a major cause of colony loss" of overwinter losses in both the 2012-2013 and the 2011/2012 surveys. Whether colony losses are attributable to CCD may be based, in part, on reported colony losses "with no dead bees present, which is indicative of CCD."[50]

Wild Bee Species

While managed bees are commercially important, wild bee species are important ecologically for sustainable forests and fields. However, no comprehensive formal statistics are available on populations of wild bees in the United States.

Of the estimated 3,500 species of solitary bees, none are thoroughly surveyed. Scientific literature is replete with assertions about the paucity of data on one group or another of native bees. Longterm data are particularly difficult to obtain for wild bees. As concluded by the researchers in a 2007 National Research Council (NRC) study, *Status of Pollinators in North America*, among all wild pollinators:

> There is evidence of decline in the abundance of some pollinators, but the strength of this evidence varies among taxa. Long-term population trends for several wild bee species (notably bumble bees) ... are notably downward. For most pollinator species, however, the paucity of longterm population data and the incomplete knowledge of even basic taxonomy and ecology make definitive assessment of status exceedingly difficult.[51]

In the absence of comprehensive data, as some of the studies cited below demonstrate, scientists seeking to find any data to analyze broad pollinator trends might be forced to rely on amalgamations of disparate studies, collections by citizen scientists,[52] and other unconventional approaches. Studies, however, show that populations of native wild pollinators are declining.[53]

"Citizen Science" Surveys of Status

As various studies demonstrate, there have been numerous attempts to promote the use of various informal "citizen science" initiatives in California, Florida, other East Coast states, and elsewhere.[54] For example, compared to most native species, bumble bees (*Bombus)* are better studied. In part this is attributable to studies which recruit citizens to gather data—for example, forming Bumble Bee Brigades in one campaign.[55] Even so, information about wild bumble bees is not comprehensive.

In 2010, an international group of scientists reviewed global trends in pollinator species, including bees.[56] They cited a study based on data gathered by citizen-scientists in the United Kingdom and in the Netherlands (number of years of observation varying by location) showing that "in both countries, bee diversity has fallen significantly in most landscapes.... Analysis of pollinator traits demonstrated that in bees ..., specialist (diet and/or habitat) and sedentary species tended to decline, whereas mobile generalists tended to thrive." As the study notes, although the "dynamics between plants and their pollinators suggest a link between the two," this is correlative and the mechanism is still not known: "It could be that plant declines are caused, in part, by a lack of pollination services, or bees could be declining owing to a lack of floral resources" or both could be declining due to "shared sensitivity to environmental changes."[57]

Another North American study examined both museum collection records and intensive nationwide surveys for records of abundance and distribution of eight bumble bee species.[58] The study showed that four species of bumble bees have declined in relative abundance by up to 96%. These four species also contracted their ranges by 23% to 87%. These changes were statistically highly significant; the other four species examined showed no clear pattern in terms of range reduction. The study hypothesized that the four declining species may be affected by a pathogen (*Nosemi bombi*) as well as reduced genetic diversity.[59] The data were adequate only to show an association, rather than a cause—were the four species declining because of the disease and low genetic diversity? Or was some other factor causing these species to decline and making them more susceptible to the disease and also reducing their genetic diversity?

Import and Release of Non-Apis Pollinating Insects

For years, conservation groups such as the Xerces Society for Invertebrate Conservation, Natural Resources Defense Council (NRDC), and Defenders of Wildlife have petitioned USDA's Animal and Plant Health Inspection Service

(APHIS) to regulate the movement of commercial bumble bees in response to concerns about population declines.[60] These groups contend that pathogens transmitted by commercial bumble bees are contributing to the spread of parasites and pathogens among wild bumble bees. Commercially reared and imported bumblebees are imported for pollination of greenhouse crops, such as tomatoes. A 2013 study by researchers in Europe showed that about 75% of bumble bee colonies imported into the United Kingdom each year are infected with parasites, and may pose risks to wild bees and honey bees.[61] The North American Plant Protection Organization (NAPPO) has published guidelines intended to assist in preparing a petition for the importation and release of non-*Apis* pollinating insects, and to assist reviewers and regulators in assessing the risk associated with their importation, movement and release. [62]

Endangered Species Act (ESA) Information

Another source of information about potential population changes among wild bee species includes listings under the Endangered Species Act (ESA, P.L. 93-205, 87 Stat. 884. 16 U.S.C. §§1531-1544). An ESA determination will include an analysis of the threats to a species, its current status and range, and other factors essential to its conservation.[63] For example, one bumble bee species (*B. franklini*) is thought to be extinct in its very limited (former) range in northeastern California and southeastern Oregon. The U.S. Fish and Wildlife Service (FWS) has this species under review for listing.[64] In addition, FWS has proposed seven species of yellow-faced bees (Genus *Hylaeus* in the Family Colletidae) for listing under ESA, with high priority for listing (2 on a scale of 12).[65] All members of this genus, including these seven, are found only in Hawaii.

In May 2013, the Xerces Society for Invertebrate Conservation and the Natural Resources Defense Council (NRDC) filed a complaint against the Department of the Interior and FWS, requesting action on a petition to grant ESA protection to the rusty patched bumble bee (*Bombus affinis*). The petitioners claim this bee was "previously common across the Upper Midwest and Eastern Seaboard, but in recent years it has been lost from 87% of its historic range." They also claim the bee is an important pollinator of prairie wildflowers, but also of cranberries, blueberries, apples, alfalfa, and other crops.[66]

Interactions with Physical Environment

Studies have showed that phenological changes[67] in flowering times are accompanied by changes in the spring emergence of bee populations. Such studies on flowering times and pollination found that 10 species of bees (including both solitary and social species) had advanced to earlier dates in the spring for their first emergence, and the flowering times of the plants they visited also advanced.[68] Advances in emergence time over the period from 1970-2010 were highly significant. The data "[support] the idea that climate change is the main factor explaining the observed phenological advances." However, the study concludes that "phenological mismatch probably has not occurred already, but that it could occur in the future, as rates of temperature warming increase." At the same time, the study noted that it analyzed generalist bees and generalist flowers—bees that visit many plant species, and plants visited by many bee species.[69] Where the life cycles of particular bee species are tied to particular plant species, different trends may occur. While no data were reported on the population sizes of these bees, the study showed no clear connection between climate change (as measured by flowering times) and any harm to the species studied. The bees seemed to be keeping up with the changes in dates of flower availability.

The phenomenon known as *trophic cascade* may also play a role in the interaction of pollinator species. Many plants can be pollinated by more than one species, and many bee species use more than one pollen or nectar source. As individual pollinator species (bees or other species) decline, other species may take over the declining species' role. Where the data are available, evidence suggests that, where a generalist bee species is able to step in to provide pollination services (even if less able than a specialist bee), the effects of a reduced bee population may be masked by the generalist. In more concrete terms, the presence of a highly efficient generalist—like the honey bee—may mask the loss of native species (regardless of the reason for the decline of the native species). If honey bees themselves then decline, the effects of the lost population may be more severe because populations of back-up pollinators have already been reduced or eliminated. And conversely, reductions in honey bee populations may reduce competition sufficiently to allow other (reduced) bee populations to rebound to greater numbers.

Another study examined population changes in meliponine (or stingless) bees. Previously stingless bees were widely cultivated by Native Americans in Central and South America before the introduction of honey bees from Europe. However, loss of forest cover eliminates nesting sites; also the shift by beekeepers to honey bees for more intensive production has also reduced the number of stingless bees. An analysis of the effects of forestry on stingless bees in Brazil observed that most bee nests were found in hollow trees over 50 cm (~20 inches) in diameter.[70] Given the high value of the bees as pollinators for many species of plants, and the low commercial value of hollow trees, the study recommended that managed forests maintain hollow trees as both seed sources and as homes for these bees. Some studies suggest that stingless bees are especially susceptible to some pesticides.[71]

FACTORS AFFECTING BEE HEALTH

Among the factors that are thought to be harming bee health are (listed in no particular order):

- *parasites, pathogens, and diseases*[72] (for a detailed listing, see text box);
- *bee genetics* including lack of genetic diversity and lineage of bees, and increased susceptibility and lowered disease resistance;
- *miticide resistance*;[73]
- *diet and nutrition* including poor nutrition due to apiary overcrowding, pollination of crops with low nutritional value, and pollen or nectar scarcity associated with invasive plants;
- *bee management issues* including transportation stress from migratory beekeeping, overcrowding, feeding practices, chemicals used by beekeepers to control mites (antibiotics and miticides), confinement and temperature fluctuations, susceptibility to disease, potential for cumulative exposure to diseases and parasites, use of bees for honey production versus pollination, chemical residue or contamination in the wax, and reliability of the queen source;
- *habitat loss, and other environmental or biological stressors* including loss of foraging area, interspecific competition between honey bees and native bees, pathogen spillover effects, and climate change;
- *pesticides* including acute or cumulative exposure to new types or combinations of agricultural pesticides through a variety of media including dust, water droplets, pollen, and nectar;
- *other agricultural practices* including the use of genetically incorporated pesticides in seeds or treated seeds, such as with bioengineered crops; and
- *potential cumulative and interactive effects* of each of these factors.

Many of these factors were among those initially thought to be contributing to CCD.[74] As outlined in USDA's 2011 progress report, the available research over the past years has led USDA and university researchers to conclude that "no single factor or specific combination of factors has been identified" as a 'cause' for CCD.[75] This has led researchers to consider CCD may be "a syndrome caused by many different factors, working in combination or synergistically";[76] it may also involve "an interaction between pathogens and other stress factors."[77]

According to USDA: "Despite a number of claims in the general and scientific media, a cause or causes of CCD have not been identified by researchers."[78] However, CCD may not be the single leading cause of honey bee colony losses in recent years.[79] USDA reports that in 2012/2013, "there were more colonies that dwindled away" rather than suffering from the onset of CCD, which is characterized by a sudden loss in bee colony populations. Whether colony loss is attributable to CCD may also be based, in part, on reported colony losses "with no dead bees present, which is indicative of CCD" (i.e., the absence of dead bees).[80] USDA also claims that "beekeepers did not report CCD as a major cause" of overwinter losses in its 2012-2013 and 2011-2012 surveys.[81]

Pests and Diseases Affecting Honey Bees

As noted in the 2007 NRC study, among the leading causes of managed honey bee losses are diseases, parasites, and recently introduced competitors. Most notable are declines due to two parasitic mites, the so-called vampire mite (*Varroa destructor*) and the tracheal mite (*Acarapis woodi*). Also of concern is the emergence of new or newly virulent fungal and viral diseases. New invasive pests are also harming bees.

Below is a listing of some identified pests and diseases. For more information, see the 2007 NRC study, *Status of Pollinators in North America*. The following is excerpted mainly from *The Beekeepers Handbook*.

Honey bee diseases may be caused by a protozoan (e.g., bacteria or amoeba), fungus, or virus.

- *Nosema disease* is the most common adult bee disease, and is caused by a microscopic fungus (formerly considered to be a protozoan). Two *Nosema* species are found in honey bees: (1) *Nosema apis*; and (2) *Nosema ceranae* (the more virulent of the two).

- *American foulbrood disease (AFD)* is the most destructive of brood diseases. (Brood refers to the egg, larval, and pupal stages in bee development.) AFD is caused by a bacterium (*Paenibacillus larvae*) that occurs in a spore or a vegetative stage; the disease is transmitted by the spore and the infected brood is killed by the vegetative stage.
- *European foulbrood disease (EFD)* is also caused by a bacterium (*Melissococcus plutonius*) and is commonly found in colonies already weakened by lack of food or by other stressors.
- *Chalkbrood disease* is caused by a fungus (*Ascophaera apis*); adult bees can detect and remove diseased larvae; honey production may be reduced but the disease usually will not destroy a colony.
- *Amoeba disease* is caused by the amoeba *Malpighamoeba mellificae* Prell; it infests the gut of honey bees. Resistant spores (cysts) form in honey bees and can transmit the disease to other bees.
- Other less common adult diseases include *septicemia* and *spiroplasma.* Both are caused by bacteria. The former cause destruction of connective tissues; the latter infects the bees' blood. They tend to cause dysentery, and arise primarily from poor food and long periods of confinement.
- Some common *viral diseases* affecting honey bees include deformed wing virus (DWV); black queen cell virus (BQCV); Israeli acute paralysis virus (IAPV); acute bee paralysis virus (ABPV); sacbrood virus (SBV); Kashmir bee virus; and chronic bee paralysis virus (CBPV).

Other pests are mostly invertebrates, but some vertebrates are problematic locally. Among these pests are parasitic mites, insects, and some larger animals.

- *Tracheal mites* (*Acarapis woodi*) are parasitic mites that live inside the breathing organs of adult bees, and eventually a newly mated female mite emerges from the old host bee.
- *Varroa mites*, now known mostly for the vampire mite (*Varroa destructor*), are large mites that feed on infested bees, resulting in disfigured, stunted adult bees and deformed larvae and pupae (varroosis).
- Emerging threats such as the parasitic *phorid fly* (*Apocephalus borealis*), known to parasitize bumble bees, have been found to also

parasitize honey bees and can eventually cause bees to abandon their hives.

- *Major insect enemies* include the wax moth and the small hive beetle.
- *Minor insect enemies* include assassin and ambush bugs, robber flies, mantids, wasps, and dragonflies.
- Various *vertebrate pests* include skunks, raccoons, bears, and mice.

One *competitor* deserves mention—the introduced Africanized honey bee. This bee is also a honey bee (*Apis mellifera*) but of a different strain from those imported from Europe. They are good foragers, but the colonies are more difficult to manage than the domestic European bee; they are sometimes called "killer bees," although an individual bee's sting is no more severe than that of a European bee. They were accidentally released in Brazil decades ago and have spread into the southern United States.

Sources: D. Sammataro and A. Avitabile, The Beekeepers Handbook, 4th ed., pp. 189-232; Beekeepers Association of Northern Virginia, "Diseases, Parasites, Pests, and Predators," presentation materials; and, for phorid flies, A. Core et al., "A New Threat to Honey Bees, the Parasitic Phorid Fly Apocephalus borealis," PLoS ONE 7(1), January 3, 2012.

Honey bee colony losses are not uncommon. The NRC study documents the extensive literature on honey bee population losses due to bee pests, parasites, pathogens, and disease.[82] Most notable are declines due to two parasitic mites, the so-called vampire mite (*Varroa destructor*) and the tracheal mite (*Acarapis woodi*), and also colony declines due to the pathogen *Paenibacillus larvae.* Other reasons for bee colony declines reported by the NRC include interspecific competition between native and introduced bees, pathogen spillover effects, habitat loss, invasive plant species that reduce nectar- and pollen-producing vegetation, bee genetics, and pesticides, among other factors.

Mite infestations are a relatively new occurrence. The 1980s saw two periods of large die-offs due to *Varroa* and tracheal mites: The first *Varroa* mite infestation was reported in 1987; tracheal mites were first detected in 1984.[83] *Varroa* mites are also said to have eliminated most feral bee colonies in the mid-1990s. *Varroa* parasitism affects both worker bees and male larvae and can affect the ability of the queen to reproduce. It is associated with viral pathogens and if left untreated can cause colony mortalities usually within six months to two years after the initial infestation. Less is known about the effects of the tracheal mite. The pathogen *Paenibacillus larvae* is the most serious honey bee pathogen and causes American foulbrood (AFB), which is a disease of larval honey bees. AFB resulted in large colony losses in the 1940s, but its incidence has been reduced by the use of antibiotics and increased apiary inspection programs. Nevertheless, mite and pathogen infestations have likely raised beekeeper operating costs to pay for miticides and/or antibiotics, labor and expenses for treatment, improved management and inspection, and colony replacement of dead bees.

According to USDA, the cost of mite controls and hive replacement every 1-2 years, as opposed to every 3-5 years previously, calls into question whether beekeeping on the United States will continue to be financially viable.[84]

Symptoms similar to those observed for CCD have also been described in the past, and heavy losses have been documented. It is still not clear whether the current colony losses are being caused by the same factors or if new contributing factors are involved. Previously, similar conditions have been termed autumn collapse, May disease, spring dwindle, disappearing disease, and fall dwindle disease.

IDENTIFIED FUTURE NEEDS

National Pollinator Health Strategy ("Presidential Memorandum")

In June 2014, the Obama Administration issued a Presidential Memorandum, "Creating a Federal Strategy to Promote the Health of Honey Bees and Other Pollinators," directing federal agencies to take steps to protect and restore domestic populations of pollinators.[85] It established a Pollinator Health Task Force, co-chaired by USDA and EPA, with representatives from the Departments of State, Defense, Interior, Housing and Urban Development, Transportation, Energy, and Education; Council on Environmental Quality; Domestic Policy Council; General Services Administration; National Science Foundation; the National Security Council Staff; Office of Management and Budget; and Office of Science and Technology Policy.

The task force is directed to develop a National Pollinator Health Strategy, which would include a Pollinator Research Action Plan to "focus federal efforts on understanding, preventing, and recovering from pollinator losses." The action plan is to be informed by studies of the health of managed honey bees and native bees; expanded collection and sharing of data related to pollinator losses; assessments of the status of native pollinators; strategies for developing affordable seed mixes; identification of existing and new methods and best practices to reduce pollinator exposure to pesticides; and strategies for targeting resources toward areas of high risk and habitat restoration potential. The strategy is to also include public education plans for individuals and businesses to help address the loss of pollinators as well as recommendations for developing public-private partnerships to build on federal efforts to encourage pollinator protection and increase the quality and amount of habitat and forage for pollinators.

The task force member agency will develop plans to increase and improve pollinator habitat. These plans may include "facility landscaping, including easements; land management; policies with respect to road and other rights-of-way; educational gardens; use of integrated vegetation and pest management; increased native vegetation; and application of pollinator-friendly best management practices and seed mixes." Member agencies are to also "evaluate permit and management practices on power line, pipeline, utility, and other rights-of-way and easements" and "make any necessary and appropriate changes to enhance pollinator habitat on federal lands through the use of integrated vegetation and pest management and pollinator-friendly best management practices" by "supplementing existing agreements and memoranda of understanding with rights-of-way holders" to establish and improve pollinator habitat.

The National Pollinator Health Strategy was scheduled to have been released in December 2014, but the report's release date has been delayed to spring 2015.[86]

In October 2014, the Council on Environmental Quality (CEQ) and the General Services Administration (GSA) issued an addendum to its sustainable landscape guidance, *Supporting the Health of Honey Bees and Other Pollinators*, to help federal agencies "incorporate pollinator friendly practices in new construction, building renovations, landscaping improvements, and in facility leasing agreements at federal facilities and on federal lands."[87] This amends previous guidance from CEQ in October 2011 (*Guidance for Federal Agencies on Sustainable Practices for Designed Landscapes*), describing strategies to achieve sustainable federal landscape practices on more than 41 million acres of land and 429,000 building assets, comprising 3.34 billion square feet of space in the United States, that are controlled or owned by the federal government.

In November 2014, USDA and EPA held a number of public listening sessions to inform the task force members. Concerns about pesticide use were among the major discussion points. Commercial beekepers and environmental groups continue to push for restrictions on pesticide use, while grower and pesticide industry groups continue to encourage broader consideration beyond pesticides.[88]

Other concerns have been raised by conservation groups and land managers regarding the possibility that commercial beekeepers will be given expanded access to public lands for foraging and habitat, which might heighten potential interspecies competition between native bees and honey bees.[89] In October 2014, USDA and other federal agencies participated in a "Honey Bee Forage and Nutrition Summit" to highlight existing policies and programs and discuss ways to improve federal programmatic efforts in establishing forage plantings for honey bees and other pollinators.[90] Representatives from USDA discussed ongoing agency efforts to improve nutrition and forage for pollinators through EQIP and the Conservation Reserve Program (CRP), among other USDA conservation programs. Other agencies described similar ongoing efforts on federal lands, including representatives from the Bureau of Land Management, the U.S. Fish and Wildlife Service, the National Park Service, and the U.S. Forest Service.

USDA-EPA Joint Report

In 2013, USDA and the U.S. Environmental Protection Agency (EPA) published a USDA-EPA Joint report, *National Stakeholders Conference on Honey Bee Health.*[91] Among the goals of the conference were to "synthesize the current state of knowledge regarding CCD, bee pests, pathogens, and nutrition, potential pesticide effects on bees, and bee biology, genetics and breeding." A summary of the key findings of the 2013 USDA-EPA Joint report, as reported by EPA, is as follows:[92]

- *Address Risks to Honey bees from Parasites and Disease.* According to the report, the parasitic *Varroa* mite is recognized as "the major factor underlying colony loss" in the United States and in other countries. Moreover, there is "widespread resistance to the chemicals beekeepers use to control mites within the hive," and new virus species have been found in the U.S. and several of these have been associated with CCD.

- *Need for Increased Genetic Diversity in Bee Colonies.* Genetic variation improves bees' thermoregulation,[93] disease resistance, and worker productivity in colonies, and bee breeding should emphasize traits (such as hygienic behavior) that confer improved resistance to *Varroa* mites and diseases.
- *Need for Improved Nutrition for Honey Bees.* Nutrition has a major impact on individual bee and colony longevity, and a poor diet can make bees more susceptible to harm from disease and parasites. Bees need better forage and a variety of plants to support colony health, and federal and state programs should consider land management strategies that maximize available nutritional forage and to protect bees by keeping them away from pesticide-treated fields.
- *Need for Collaboration and Information Sharing.* Best Management Practices (BMPs) associated with pesticide use and bees are known but are not widely or systematically followed by U.S. crop producers. "Informed and coordinated communication between growers and beekeepers" is needed, along with "effective collaboration between stakeholders on practices to protect bees from pesticides." Beekeepers have identified the need for "accurate and timely bee kill incident reporting, monitoring, and enforcement."
- *Additional Pesticide Research Needed.* According to EPA: "The most pressing pesticide research questions relate to determining actual pesticide exposures and effects of pesticides on bees in the field and the potential for impacts on bee health and productivity of whole honey bee colonies."

The National Honey Bee Health Stakeholder Conference Steering Committee is made up of representatives from Pennsylvania State University; from USDA's Office of Pest Management Policy (OPMP), National Institute of Food and Agriculture (NIFA), Agricultural Research Service (ARS), Animal and Plant Health Inspection Service (APHIS), Natural Resources Conservation Service (NRCS), and National Agricultural Statistics Service (NASS); and from EPA's Office of Pesticide Programs (OPP).

Much of the current research on bee health is being conducted by scientists at USDA and its Beltsville bee laboratory, the USDA-supported Bee Informed Partnership, and scientists at many of the land-grant universities nationwide.

USDA Actions and Funding

CCD Action Plan

USDA released its initial action plan for addressing CCD in July 2007. USDA's action plan focuses on improving coordination and redirecting existing resources and research for mitigation and prevention, including education and outreach, as well as expanding research and diagnostic resources to prevent future losses, working with the land grant universities. It also coordinates activities across three USDA agencies: Agricultural Research Service (ARS), Animal and Plant Health Inspection Service (APHIS), and National Institute of Food and Agriculture (NIFA).

Under the plan, USDA would (1) conduct surveys and collect data on bee health; (2) analyze bee samples for pests, disease-causing pathogens, pesticide exposure, and other factors; (3) conduct controlled experiments to identify factors affecting bee health, including potential causes of colony collapses; and (4) develop best management practices and guidelines to improve general bee health and reduce susceptibility to colony collapses and other disorders among both honey bees and non-*Apis* bees.[94] Since 2009, USDA has published a series of annual progress reports on its CCD research.[95] The progress reports provide detailed information on the status of ongoing research under each of the four elements of USDA's action plan, including survey and (sample) data collection, analysis of existing samples, research to identify factors affecting honey bee health, and mitigative and preventive measures.

Funding to Address Pollinator Concerns

Funding for honey bee and pollinator research at USDA's ARS has increased since the enactment of the 2008 farm bill (P.L. 110-246) and in subsequent appropriations, which, among other things, provide additional funding for research and conservation programs addressing honey bees and pollinators. Total ARS funding for honey bee research has been as follows:[96]

- FY2007 - $7,675,000
- FY2008 - $7,798,000
- FY2009 - $8,290,000
- FY2010 - $10,000,000
- FY2011 - $9,980,000 (reduction in funding was due to the FY2011 rescission)
- FY2012 - $10,128,000

- FY2013 - $9,353,000 (reduction in funding was due to sequestration and rescission); and
- FY2014 - $10,309,000.

Of these annual amounts, about $3 million is spent to address the *Varroa* mite.[97] ARS also has an "Area-wide Project on Bee Health," which consists of temporary funding of about $1.3 million in FY2013. Additional funding is available to USDA's NIFA, and includes combined research on honey bees, funding specific to CCD and bee and pollinator health, including funding for various research labs and grants. Emerging issues grants have been awarded to Penn State University and the University of Georgia to study the effects of pesticides, pathogens, and miticides on pollinator populations.[98]

Additional funding has been allocated in conjunction with the release of the June 2014 Presidential Memorandum. For example, USDA's Natural Resources Conservation Service (NRCS) awarded another $4 million in technical and financial assistance to farmers and ranchers to improve the health of honey focusing on crop production in five midwestern states: Michigan, Minnesota, North Dakota, South Dakota, and Wisconsin.[99] USDA's Conservation Reserve Program (CRP) also provided $8 million in incentives for Michigan, Minnesota, North Dakota, South Dakota, and Wisconsin farmers and ranchers who establish new habitats for declining honey bee populations.[100] NIFA also provided a $6.9 million grant to Michigan State University to develop sustainable pollination strategies for U.S. specialty crops funded through the Specialty Crop Research Initiative.[101] This additional funding is part of the Administration's FY2015 proposal to provide up to $70 million to focus broadly on pollinator declines, which includes a proposed initiative to create a $25 million competitive program administered by NIFA ("Innovative Institute on Pollination and Pollinator Health").[102]

Some beekeepers will also receive disaster assistance aid through the USDA's Emergency Assistance for Livestock, Honeybees, and Farm-Raised Fish Program (ELAP). (For more information, see "Insurance and Disaster Provisions.")

Some states and/or state beekeeper associations are also providing grant money to new and existing beekeepers to purchase new hives or additional equipment and beekeeping supplies, or to maintain local queen breeding groups that support local beekeeping communities.[103]

In addition, in January 2014, EPA announced it had awarded nearly $500,000 in agricultural grants for Integrated Pest Management (IPM) practices to reduce the use of potentially harmful pesticides and lower risk to bees. The IPM grants were awarded to Louisiana State and Penn State University, and University of Vermont.[104]

Issues for Congress

Following concerns over honey bee colony losses in 2006-2007, Congress provided for increased funding for bee research, among other types of farm program support to protect pollinators, as part of the 2008 farm bill. Given continued concerns about the health and well-being of honey bees and other pollinators, this issue has continued to be legislatively active in Congress, and various proposals were again considered as part of farm bill debates in both the House and Senate in 2013. For the 114th Congress, the incoming chairman of the House Agriculture Subcommittee on Biotechnology, Horticulture, and Research has said that bee health will be a top issue for his subcommittee.[105] In addition, various other legislative proposals have been considered outside the farm bill debate that could broadly support other types of bee and pollinator protections and habitat areas.

Farm Bill Provisions

The Agricultural Act of 2014 (P.L. 113-79, H.Rept. 113-333), referred to here as the "2014 farm bill," is the most recent omnibus farm bill. It was enacted in February 2014 and succeeded the Food, Conservation, and Energy Act of 2008 (P.L. 110-246, "2008 farm bill").[106] The 2014 farm bill reauthorized and expanded many of the 2008 farm bill provisions that address honey bees and pollinators as part of the law's research, conservation, specialty crop, and miscellaneous title provisions. Overall, regarding managed honey bees and native pollinators, the 2014 farm bill:

- extends through 2018 various USDA research and conservation provisions directed at protecting pollinators, enacted in the 2008 farm bill (see "Research Provisions" and "Conservation Provisions," below, for more information);
- expands annual reporting requirements on CCD along with other health disorders;

- requires USDA to publish guidance on enhancing long-term pollinator health and long term viability of pollinators, including addressing habitat and forage needs for native pollinators and managed honey bees; and
- requires USDA to assess federal efforts to mitigate pollinator losses and threats to the U.S. commercial beekeeping industry, and to recommend how to better coordinate federal agency efforts to address the decline of managed honey bees and native pollinators.

Some proposed provisions were not included in the 2014 farm bill. In particular, a House-passed provision, Protection of Honey Bees and Other Pollinators,[107] would have required USDA, in consultations with the Department of the Interior and EPA, to take measures to improve federal coordination in addressing the documented decline of managed and native pollinators and promote the long-term viability of honey bee, wild bees and other beneficial insects in agriculture. The provision also would have required USDA to establish a federal task force on bee health and commercial beekeeping to coordinate and assess efforts to mitigate pollinator losses, and the task force is to submit a report to Congress within 180 days of enactment of the farm bill, as well as conduct a study to consider relocating and modernizing pollinator research labs.

Research Provisions

The research title of the 2008 farm bill identified pollinator protection among its so-called high-priority research and extension areas (P.L. 110-246, §7204). It provided for research and extension grants (1) to survey and collect data on bee colony production and health; (2) to investigate pollinator biology, immunology, ecology, genomics, and bioinformatics; (3) to conduct research on various factors that may be contributing to or associated with colony collapse disorder and other serious threats to the health of honey bees and other pollinators, including parasites and pathogens of pollinators, and the sublethal effects of insecticides, herbicides, and fungicides on honey bees and native and managed pollinators; (4) to develop mitigative and preventative measures to improve native and managed pollinator health; and (5) to promote the health of honey bees and native pollinators through habitat conservation and best management practices.[108] The 2008 farm bill authorized appropriations for grants at $10 million annually for FY2008-FY2013.

The research provisions also directed USDA to increase its capacity and infrastructure to address colony collapse disorder and other long-term threats to pollinator health (including hiring additional personnel) and to conduct research on colony collapse disorder and other pollinator issues at USDA's facilities. As amended, annual appropriations were authorized at $7.25 million (FY2008-FY2013), with another $2.75 million annually (FY2008-FY2013) for honey bee pest and pathogen surveillance. This authorized funding has supported honey bee and pollinator research efforts at USDA's ARS in recent years (see "Funding").

The 2008 farm bill also directed USDA to submit an annual report to Congress on its response to CCD, indicating that the report should investigate the cause(s) of honey bee colony collapse and recommend appropriate strategies to reduce colony loss (see "CCD Action Plan").

The 2014 farm bill reiterated that USDA should continue to focus its research efforts on honey bees and native pollinators (P.L. 113-79, §7209). First, the farm bill reauthorized and expanded the "high-priority research and extension initiatives" at USDA regarding honey bees, including the requirement that USDA conduct a nationwide "honey bee pest, pathogen, health, and population status surveillance" program, among other efforts intended to counteract the declining honey bee health and declining populations of honey bees and native pollinators. Second, the farm bill reauthorized and expanded on the requirement that USDA submit an annual report describing honey bee colony collapse and honey bee health disorders, and find appropriate strategies, including best management practices (BMPs) to reduce colony losses, and address the decline of managed honey bees and native pollinators. USDA's report must also assess federal efforts to mitigate pollinator losses and threats to the United States commercial beekeeping industry and recommend to Congress how to better coordinate federal agency efforts.

The 2014 farm bill provision in the Research title extended each of the authorized appropriations in the 2008 farm bill.[109] Total authorized appropriations are $20 million annually through FY2018, and include

- research and extension activities ($10 million annually);
- USDA capacity and infrastructure ($7.25 million annually); and
- honey bee surveillance ($2.75 million annually).

Third, the farm bill directed USDA to consult with the Department of the Interior and EPA, and publish "guidance on enhancing pollinator health and the long-term viability of populations of pollinators," including recommendations related to allowing for managed honey bees to forage on

National Forest System (NFS) lands. Under current law, honey bees can be allowed on NFS lands under a "special use authorization permit."[110]

Conservation Provisions

The conservation title of the 2008 farm bill included language that broadly encourages habitat development and protection among the administrative requirements for native and managed pollinators under USDA's conservation programs (P.L. 110-246, §2708), and ensures that USDA's conservation technical assistance includes standards that account for native and managed pollinators (P.L. 110-246, §2706). These provisions could broaden the focus of USDA's farm conservation programs to include pollinator habitats and habitat improvement among their goals, as well as require USDA to review its conservation practice standards with respect to managed and native pollinators.

The 2014 farm bill generally continued these types of protections. In February 2014, USDA's Natural Resources Conservation Service (NRCS) announced that it would provide nearly $3 million in technical and financial assistance for farmers and ranchers through the Environmental Quality Incentives Program (EQIP) to help promote the health of honey bees by implementing conservation practices that provide bees with forage areas while providing other environmental benefits. This funding is targeted in five midwestern states: Michigan, Minnesota, North Dakota, South Dakota, and Wisconsin.[111] In October 2014, USDA will be providing more than $4 million in technical and financial assistance to farmers and ranchers in these states to improve honey bee health. USDA notes that this effort contributes to the efforts of the Presidential Memorandum to create a federal strategy to promote the health of honey bees and other pollinators.[112]

The need for habitat and forage areas for native pollinators and managed honey bees—to supply pollen and nectar to address the nutritional needs of bees and other pollinators—continues to be an important policy issue. Some advocacy groups asked USDA to convene a "Honey Bee Nutrition and Forage Summit" to examine the importance of nutrition and forage to the health of honey bees and sustainable agricultural pollination services,[113] which USDA hosted in October 2014. Forage efforts are also being considered in some state legislatures including in California to help address pollination service needs. USDA is also focusing on forage areas as part of its conservation programs; see "National Pollinator Health Strategy ("Presidential Memorandum")."

Insurance and Disaster Provisions

Other provisions in the 2008 farm bill supported pollinators through the bill's crop insurance and other disaster assistance provisions. One such provision identifies honey farms as possible beneficiaries of the bill's supplemental agricultural disaster assistance (P.L. 110-246, §12033); another provision provides contracts for additional policies and studies to carry out research and development regarding insurance policies that cover loss of bees (P.L. 110-246, §12023).

As authorized by the 2008 farm bill, USDA established the Emergency Assistance for Livestock, Honey Bees, and Farm-Raised Fish Program (ELAP).[114] This program, administered by USDA's Farm Service Agency (FSA), provides emergency relief to producers of livestock, honey bees, and farm-raised fish to aid in the reduction of losses caused by disease, adverse weather, or other natural disaster conditions, such as blizzards and wildfires. Eligible honey bee producers—those who incur physical losses of honey bees and honey bee hives because of colony collapse disorder—must provide documentation, and/or a certification that the loss of honey bees was due to CCD, from one or more of the following: registered entomologist; cooperative extension specialist; and/or land grant university. Supplemental agricultural disaster assistance under ELAP initially covered certain losses that occurred on or after January 1, 2008, and before October 1, 2011. ELAP expired in 2011.

The 2014 farm bill reinstated ELAP retroactively to a period covering October 2011 through September 2018, providing up to $20 million annually in mandatory funding to compensate producers for disaster losses not covered under other disaster programs (P.L. 113-79, §1501). Beekeepers with dead bees may be eligible for payments under ELAP. Among eligible causes of bee loss are those that are a "direct result of an eligible adverse weather or loss condition, including but not limited to, colony collapse disorder (CCD) (colony loss only), earthquake, eligible winter storm (colony loss only), excessive wind, flood, hurricane, lightning, tornado, volcanic eruption or wildfire natural disasters as well as colony collapse disorder."[115] Eligible honeybees under ELAP include "bees housed in a managed hive and used for honey production, pollination or honeybee breeding" and do not include "wild, feral honeybees, leaf cutter bees or other bee species that are not used for producing honey, pollinating or breeding honeybees." Under ELAP, FSA provides payments for honey bee colony or hive losses:

- ELAP payments for honey bee *colony losses* are based on a minimum of 75% of the number of honeybee colonies lost in excess of normal mortality (17.5%)[116] due to an eligible adverse weather or loss condition, multiplied by the average fair market value per honeybee colony for the applicable program year.
- ELAP payments for honey bee *hive losses* are based on a minimum of 75% of the number of honeybee hives lost due to an eligible adverse weather or loss condition, multiplied by the average fair market value per honeybee hive for the applicable program year.

For the 2012-2014 program years, the FSA-established average fair market values ranged from $75 to $85 per colony, and ranged from $210 to $230 per hive. ELAP also compensates eligible beekeepers based on a minimum of 60% of the actual cost of purchased or harvested feed that was intended for honey bees but damaged due to an eligible adverse weather or loss condition. Certain payment eligibility and limits may apply.

In November 2014, USDA announced that nearly 2,500 applicants will receive disaster assistance aid through ELAP for losses suffered from October 1, 2011, through September 30, 2013.[117]

Other Farm Bill Provisions

The 2014 farm bill also requires that USDA submit to the Food and Drug Administration a report that describes an appropriate federal standard for the identity of honey, including consideration of the March 2006 Standard of Identity citizens petition filed with FDA (P.L. 113-79, §10012). The agency published a notice to solicit public comments on "how a federal standard of identity for honey would be in the interest of consumers, the honey industry, and U.S. agriculture."[118] ("Standard of Identity" broadly refers to establishing definitions and standards for commodities such as honey.)

The 2008 farm bill also contained provisions that generally support honey production. These include, for example, provisions pertaining to the National Honey Board (P.L. 110-246, §§10401- 10402); provisions covering rates for marketing assistance loans for certain commodities, including honey (P.L. 110-246, §1202); and provisions covering certain nutrition title provisions (such as P.L. 110-246, §4231).

Other Legislative Proposals

Additional legislative proposals regarding honey bees and other pollinators have been offered outside of the farm bill debate.

In the 113th Congress, Representative Alcee Hastings introduced H.R. 4790, Highways Bettering the Economy and Environment Act (or Highways BEE Act). The bill would promote conservation practices on 17 million acres of highway "rights-of-ways" (managed by state Departments of Transportation or state DOTs), and would promote native plantings that provide improved habitat, forage, and migratory corridors for pollinators, ground nesting birds, and other small wildlife, as well as encourage reduced mowing and overall maintenance costs to state DOTs. Roadsides are also recognized as a potential habitat for native wildlife, including pollinating insects.[119] The proposed bill language was not included as part of broader transportation legislation considered during the 112th Congress. The bill was also introduced in the 112th Congress as H.R. 2381, and has been actively promoted by Pollinator Partnership/North American Pollinator Protection Campaign (NAPPC).[120]

In the 113th Congress, Representatives Earl Blumenauer and John Conyers, Jr. introduced H.R. 2692, Saving America's Pollinators Act of 2013. The bill would suspend registrations of neonicotinoids and ban new registrations of any pesticide for use on "bee attractive plants, trees, and cereals" until the EPA determines that the insecticide will not cause "unreasonable adverse effects" on pollinators, including native bees, honey bees, and other beneficial insects, as well as birds and bats. The bill also would require the Department of the Interior to coordinate with EPA in monitoring the health and populations of native bees, and annually report to Congress on their health and population status. This bill is in part a response to reports that 50,000 bees were found dead in a suburban shopping-center parking lot in Oregon in June 2013, reportedly due to exposure to pesticides used on trees near the parking lot to control aphids.[121]

Also in the 113th Congress, Representative Austin Scott introduced H.R. 5447, which would amend U.S. pesticide laws to provide for expedited registration of pesticides that "improve managed pollinator bee health, including managing resistance to parasitic pests," and for expedited application for a pesticide registration that is "reasonably expected to improve the health of managed pollinator bees, including managing resistance to parasitic pests of managed pollinator bees." The bill would also require USDA and EPA to submit reports to Congress on the "extent and scope of the threat to the health of managed pollinator bees" and the availability of pesticides to manage bee pests, particularly the *Varroa* mite.

End Notes

[1] N. W. Calderone, "Insect Pollinated Crops, Insect Pollinators and US Agriculture: Trend Analysis of Aggregate Data for the Period 1992–2009," PLoS ONE 7(5): e37235, May 2012. Updated values are for 2010.

[2] Ibid.

[3] United Nations Environment Programme (UNEP), Global Honey Bee Colony Disorders and Other Threats to Insect Pollinators, UNEP Emerging Issues, 2010.

[4] For more information, see CRS Report RL33938, Honey Bee Colony Collapse Disorder.

[5] D. vanEngelsdorp et al., "Preliminary Results: Honey Bee Colony Losses in the United States, Winter 2012-2013," http://beeinformed.org/2013/05/winter-loss-survey-2012-2013/.

[6] K. Kaplan, "Fact Sheet: Survey of Bee Losses during Winter of 2012/2013," May 2013, http://www.ars.usda.gov/is/ br/beelosses/index.htm.

[7] For more information, see CRS Report R43076, The 2014 Farm Bill (P.L. 113-79): Summary and Side-by-Side.

[8] National Academy of Sciences, National Research Council (NRC), Status of Pollinators in North America, 2007, p. 36; and S. Fecht, "Hive and Seek: Domestic Honeybees Keep Disappearing, but Are Their Wild Cousins in Trouble, Too?" Scientific American, May 8, 2012.

[9] National Honey Board, "Beekeepers and Honeybee Colonies," http://www.honey.com/ nhb/media/press-kit/. Data from 2012 industry survey, Bee Culture magazine.

[10] A colony of bees is composed of between 250 and 50,000 individual bees (D. Sammataro and A. Avitabile, The Beekeepers Handbook, 4th ed.). Most larger healthy managed hives have between 20,000 and 50,000 bees.

[11] These operations are able to supply honey bee colonies during the critical phase of a crop's bloom cycle, when bees pollinate a crop as they fly from flower to flower collecting nectar and pollen, which they carry back to the nest.

[12] The shift by beekeepers to honey bees for more intensive production has also reduced the number of stingless bees.

[13] Iowa State University, "Native Bees of North America," http://bugguide.net/ node/ view/475348.

[14] NRC, Status of Pollinators in North America, 2007, p. 50; and Xerces Society, "Pollinator Conservation: Native Bee Biology," http://www.xerces.org/native-bees/.

[15] NRC, Status of Pollinators in North America, 2007, p. 43.

[16] Refers to bees that lay their eggs in another bee's nest; as their eggs hatch early, the cuckoo larvae eat the other's provisions.

[17] N. Gallai et al., "Economic valuation of the vulnerability of world agriculture confronted with pollinator decline," Ecological Economics, 2009, 68: 810-821. Converted from the reported estimate of €153 billion (value in 2005).

[18] N. W. Calderone, "Insect Pollinated Crops, Insect Pollinators and US Agriculture: Trend Analysis of Aggregate Data for the Period 1992–2009," PLoS ONE 7(5): e37235, May 2012. Updated values are for 2010.

[19] M. R. Berenbaum, University of Illinois, statement before the Subcommittee on Horticulture and Organic Agriculture, U.S. House of Representatives, March 29, 2007; and J. Pettis, USDA's ARS, interview with University of Pennsylvania staff, January 23, 2007.

[20] Staple crops (wheat, corn, and rice) do not rely on insect pollination and are mostly wind-pollinated.

[21] R. A. Morse and N. W. Calderone, The Value of Honey Bees as Pollinators of U.S. Crops in 2000, March 2000, Cornell University; and A. M. Klein et al., "Importance of pollinators in changing landscapes for world crops," Proceedings of the Royal Society B: Biological Sciences, Vol. 274, No. 1608, February 7, 2007.

[22] Crops include apricot, citrus, peaches, pears, nectarines, plums, grapes, brambleberries, strawberries, olives, melon (cantaloupe, watermelon, and honeydew), peanuts, cotton, soybeans, and sugarbeets.

[23] USDA, "Agroforestry: Sustaining Native Bee Habitat Crop Pollination," AF note-32, August 2006.

[24] See February 2014 edition of Journal of Pollination Ecology (www.pollinationecology.org). See also the UN's website (http://www.internationalpollinatorsinitiative.org/).

[25] L.A. Garibaldi, et al., "Wild Pollinators Enhance Fruit Set of Crops Regardless of Honey Bee Abundance," Science, Volume 339: 1608-1611, March 29, 2013; H. Hamilton, "Buzz On Native Bees," USGS blog posting, June 17, 2013; and S. Milius, "Native Pollinators Boost Crop Yields Worldwide," ScienceNews, March 1, 2013.

[26] See, for example, M. A. Aizen and L. D. Harder, "The Global Stock of Domesticated Honey Bees Is Growing Slower than Agricultural Demand for Pollination," Current Biology, May 2009.

[27] N. W. Calderone, "Insect Pollinated Crops, Insect Pollinators and US Agriculture: Trend Analysis of Aggregate Data for the Period 1992–2009," PLoS ONE 7(5): e37235, May 2012. Other available estimates for the total value of bee and insect pollination range from $5.7 billion to $19.0 billion annually (as reported by the NRC in 2007).

[28] Other known insect pollinators are wasps, hover flies, other flies, beetles, thrips, ants, butterflies, and moths.

[29] R. Chaplin-Kramer et al., "Value of Wildland Habitat for Supplying Pollination Services to Californian Agriculture," Rangelands, June 2011, Vol. 33(3): 33-41.

[30] A.V. Smith, "Where the Wild Bees Roam Native pollinators need more protection," EugeneWeekly, September 11, 2014. See also S. Rao and W. P. Stephen, "Abundance and Diversity of Native Bumble Bees Associated with Agricultural Crops: The Willamette Valley Experience," Psyche, Volume 2010, December 2009.

[31] Some "spillover" pollination occurs, including pollination from colonies owned by part-time beekeepers and hobbyists, or pollination of adjacent fields from commercial hives.

[32] Almond Board of California, http://www.almonds.com/growers/pollination.

[33] "Honey Bees and Blueberry Pollination," Fact Sheet 629, University of Maine Extension, April 2002.

[34] T.D. Breeze, et al., "Agricultural Practices Exacerbate Honeybee Pollination Service Supply-Demand Mismatches Across Europe," PLoS ONE 9(1): e82996, January 2014.

[35] D. M. Caron, R. Sagili, and M. Cooper, "Pacific Northwest (PNW) 2011 Beekeeper Pollination."

[36] D. M. Caron, "Bee Colony Pollination Rental Prices, Eastern U.S. with Comparison to West Coast" (Table 1, Crop Pollination in MAAREC Region (PA, DE, NJ, MD, VA, WV) and CA, 2008-2009), 2011.

[37] NRC, Status of Pollinators in North America, 2007, p. 3.

[38] Ibid.

[39] USDA, 2012 Census of Agriculture, Table 21 (state-level data). Honey sales were valued at $264 million in 2012.

[40] Presentation to Congressional staff by Jeff Pettis, USDA/Agricultural Research Service, June 18, 2012.

[41] USDA, Honey, March 2014, http://usda.mannlib.cornell.edu/usda/current/Hone/Hone-03-21-2014.pdf.

[42] The Bee Informed Partnership is an extension project that seeks to decrease the number of honey bee colonies that die over the winter, in part through the creation of the so-called Bee Informed Database of honey bee health. This database will contain data collected by the Bee Informed Partnership survey efforts, as well as data from other honey bee health monitoring projects. Funding for the survey is from the Agriculture and Food Research Initiative of USDA's National Institute of Food and Agriculture. For more information, see http://beeinformed.org/about/bip-database/.

[43] For more information, see the Bee Informed Partnership website: http://beeinformed.org/.

[44] Pollinator Partnership, "Helping the earth by supporting pollinators," 2009 publication. Estimates attributed to Jeffrey S. Pettis at USDA's Agricultural Research Service (ARS).

[45] D. vanEngelsdorp et al., "Colony Loss 2013-2014," May 15, 2014, http://beeinformed.org/2014/05/colony-loss2013-2014/. Past surveys are all posted on the Bee Informed Partnership website. To maintain a strong, healthy colony that will winter well and keep a large bee population through the following spring, the colony needs to rear a large number of bees capable of overwintering.

[46] K. Kaplan, "Fact Sheet: Survey of Bee Losses during Winter of 2012/2013," May 2013.

[47] D. vanEngelsdorp, R. Underwood, D. Caron, and J. Hayes Jr., "An Estimate of Managed Colony Losses in the Winter of 2006-2007: A Report Commissioned by the Apiary Inspectors of America," American Bee Journal, July 2007. Based on a survey of beekeepers that included 384 respondents representing 153,000 managed bee colonies located in AR, FL, GA, MD, MI, MS, MT, NM, ND, OH, PA, SD, TN, and WI.

[48] These statistics may have been misrepresented in the popular press, which often state that 25% of the nation's 2.4 million colonies have been lost (citing the AIA survey as its source).

[49] K. Kaplan, "Fact Sheet: Survey of Bee Losses during Winter of 2012/2013," May 2013, http://www.ars.usda.gov/is/ br/beelosses/index.htm.

[50] USDA, CCD Progress Report, June 2012, http://www.ars.usda.gov/is/br/ccd/ ccdprogressreport2012.pdf, p. 9.

[51] For more information, see CRS Report RL31654, The Endangered Species Act: A Primer.

[52] A well-known example of citizen-scientist-based data in North America is the annual Christmas bird count, which has been conducted for nearly a century in some areas. Data may not serve for some quantitative analyses, but may be much more reliable for the presence or absence of a bird, for example. In other studies, on arrival dates at bird feeders, or bees on spring flowers, the date on which a species is first observed could be another relatively reliable observation.

[53] See, for example, L.A. Burke, et al., "Plant-Pollinator Interaction over 120 Years: Loss of Species, Co-Occurrence, and Function," Science, Volume 339: 1611-1615, March 29, 2013.

[54] See, for example, the University of California's backyard bee count (http://www.greatsunflower.org/), the University of Florida's "Native Buzz" project (http://www.ufnativebuzz.com/), and the "Bee Hunt" project organized by multiple university staff, federal agencies, and private partnerships (http://www. discoverlife.org/bee/index.html).

[55] The University of Wyoming, through its Berry Center, sponsors the program. See http://www.uwyo.edu/berrycenter/ citizen-science/bumblebees.html for more information.

[56] S. G. Potts, J. C. Biesmeijer, C. Kremen et al., 2010, "Global pollinator declines: trends, impacts and drivers," Trends in Ecology and Evolution, v. 25, n. 6, pp. 345-353.

[57] Ibid.

[58] S. A. Cameron, J. D. Lozier, J. P. Strange et al., 2011, "Patterns of widespread decline in North American bumble bees," Proceedings of the National Academy of Sciences, v. 108, n. 2, pp. 662-667.

[59] The study notes that some believe Nosema bombi was introduced via commercially imported European bumble bees, and North American bumble bees may therefore have less resistance to the disease than European species.

[60] Xerces press release, "Protection for Pollinators: Conservation Groups and Scientists Push USDA to Save Wild Bumble Bees," October 29, 2013. See also North American Pollinator Protection Campaign (NAPPC), Importation of Non-Native Bumble Bees into North America: Potential Consequences of Using Bombus terrestris and Other Non-Native Bumble Bees for Greenhouse Crop Pollination in Canada, Mexico, and the United States, August 2006.

[61] P. Graystock et al., "The Trojan hives: pollinator pathogens, imported and distributed in bumblebee colonies," Journal of Applied Ecology, Volume 50, Issue 5: 1207–1215, October 2013.

[62] NAPPO, "Guidelines for the Petition for Import and Release of Non-Apis Pollinating Insects into NAPPO Countries," NAPPO Regional Standards for Phytosanitary Measures (RSPM) No. 29, October 2008.

[63] NRC, Status of Pollinators in North America, 2007, p. 7. See also p. 88 (referring to Bombus, or bumble bees); p. 29 (referring to pollinators generally in North America); and p. 203 (referring to the absence of baseline data on pollinator status generally as an impediment to estimation of any decline).

[64] See FWS, "Species Profile," http://ecos.fws.gov/speciesProfile/ profile/species Profile.action?spcode=I0IR.

[65] See FWS, "Endangered Species," http://www.fws.gov/endangered/.

[66] Xerces press release, "Rusty Patched Bumble Bee Threatened with Extinction," May 13, 2014. The complaint is available at http://www.xerces.org/wp-content/uploads/2014/05/ Bee-Case-Complaint-as-filed.pdf.

[67] Phenology refers to the scientific study of biological phenomena that are cyclical or periodic. Examples include migration, hibernation, and nesting, as well as flowering times, and is important in analysis of climate change.

[68] See, for example, I. Bartomeus, et al., 2011, "Climate-associated phenological advances in bee pollinators and bee-pollinated plants," Proceedings of the National Academy of Sciences, v. 108, n. 51, pp. 20645-20649.

[69] Across four genera: Andrena (three species), Bombus (two species), Colletes (one species), and Osmia (four species).

[70] G. C. Venturieri, 2009, "The impact of forest exploitation on Amazonian stingless bees (Apidae Meliponini)," Genetics and Molecular Research, v. 8, n. 2, pp. 684-689.

[71] H. V. V. Tomé, G. F. Martins et al., 2012, "Imidacloprid-induced impairment of mushroom bodies and behavior of the native stingless bee Melipona quadrifasciata anthidioides," PLoS ONE, v. 7, n. 6, p. e38406.

[72] Among the leading causes of declining bee populations cited in the 2007 NRC study—which was published before CCD became a well-publicized phenomenon—were pathogens and

introduced parasites, particularly Varroa destructor, the varroa mite. That study, among others, documents the extensive, but still inconclusive literature on honey bee population losses due to bee pests, parasites, pathogens, and disease, as well as other causes.

[73] A miticide is a pesticide intended to kill mites.

[74] For more information, see CRS Report RL33938, Honey Bee Colony Collapse Disorder.

[75] USDA, CCD Progress Report, June 2011, http://www.ars.usda.gov/is/br/ccd/ ccdprogress report2011.pdf.

[76] USDA, CCD Progress Report, June 2009, http://www.ars.usda.gov/is/br/ccd/ ccdprogress report.pdf.

[77] D. vanEngelsdorp et al., 2009, "Colony collapse disorder: A descriptive study," PLOS One, v. 4, n. 8, (August).

[78] USDA, "Honey Bees and Colony Collapse Disorder," http://www.ars.usda.gov/News/ docs.htm?docid=15572.

[79] K. Kaplan, "Fact Sheet: Survey of Bee Losses During Winter of 2012/2013."

[80] USDA, CCD Progress Report, June 2012, http://www.ars.usda.gov/is/br/ccd/ ccdprogress report2012.pdf, p. 9.

[81] K. Kaplan, "Fact Sheet: Survey of Bee Losses During Winter of 2012/2013."

[82] National Academy of Sciences, National Research Council, Status of Pollinators in North America, 2006.

[83] Interview with Maryann Frazier, Senior Extension Agent, Pennsylvania State University, January 28, 2007.

[84] Testimony of Jeff Pettis, USDA, at a House Agriculture Subcommittee on Horticulture, Research, Biotechnology, and Foreign Agriculture hearing, April 29, 2014.

[85] The White House, "Presidential Memorandum—Creating a Federal Strategy to Promote the Health of Honey Bees and Other Pollinators," June 20, 2014, http://www.whitehouse.gov/ the-press-office/2014/06/20/presidentialmemorandum-creating-federal-strategy-promote-health-honey-b. For general information on presidential memoranda, see CRS Report RS20846, Executive Orders: Issuance, Modification, and Revocation.

[86] Announcement by Michael Stebbins, White House Office of Science and Technology Policy, at the 14th annual conference of the North American Pollinator Protection Campaign (NAPPC), October 22, 2014.

[87] CEQ, "Guidance for Federal Agencies on Sustainable Practices for Designed Landscapes and Supporting Pollinators on Federal Landscapes," http://www.whitehouse.gov/administration/ eop/ceq/sustainability/landscaping-guidance and http://www.whitehouse.gov/ sites/default/ files/docs/supporting_the_health_of_honey_bees_and_other_pollinators.pdf. Involvement from other federal agencies included representatives from USDA and the U.S. Forest Service, the Smithsonian Institution, U.S. Botanic Garden, U.S. Geological Survey, the National Science Foundation, and the Departments of Transportation, Defense, Education, Veterans Affairs, and the Department of Interior's Office of Planning and Management, with support from the White House Office of Science and Technology Policy.

[88] J.R. Pegg, "Neonicotinoid Controversy Hits Pollinator Health," Food Chemical News, December 3, 2014.

[89] See, for example, T. Stecker, "Policy Fight Looms as Agencies Treat honeybee—a 17th-century import—as Exotic Invader," Energy and Environmental Policy News, December 2, 2014.

[90] Comments by USDA representatives, "Honey Bee Forage and Nutrition Summit," October 20, Alexandria VA.

[91] USDA, Report on the National Stakeholders Conference on Honey Bee Health, National Honey Bee Health Stakeholder Conference Steering Committee, May 2013, http://www.usda.gov/documents/ReportHoneyBeeHealth.pdf.

[92] EPA news release, "USDA and EPA Release New Report on Honey Bee Health," May 2, 2013.

[93] Refers to the ability to keep body temperature steady even if the surrounding environment is different.

[94] USDA, CCD Steering Committee, "Colony Collapse Disorder Action Plan," June 20, 2007, http://www.ars.usda.gov/ is/br/ccd/ccd_actionplan.pdf.

[95] USDA, Colony Collapse Disorder Progress Report, various years, available at http://www.ars.usda.gov/News/ docs.htm?docid=15572.

[96] CRS communication with USDA personnel, December 18, 2009.

[97] Testimony of Jeff Pettis, USDA, at a House Agriculture Subcommittee on Horticulture, Research, Biotechnology, and Foreign Agriculture hearing, April 29, 2014.

[98] Statements by Jeff Pettis, USDA, to Senate Environment and Public Works committee staff, April 9, 2008.

[99] USDA press release, "USDA to Provide $4 million For Honey Bee Habitat," October 29, 2014.

[100] USDA press release, "USDA Provides $8 Million to Help Boost Declining Honey Bee Population," June 20, 2014.

[101] USDA press release, "USDA Awards Grant to Michigan State University to Study Pollination Solutions for Specialty Crops," August 6, 2014. Other funding includes $45 million in combined mandatory and discretionary funds for USDA and EPA (see USDA, FY2015 Budget Summary and Annual Performance Plan, April 2014).

[102] USDA, NIFA, "2015 Explanatory Notes," http://www.obpa.usda.gov/19nifa2015notes.pdf, p. 19-38.

[103] See, for example, Virginia's Beehive Grant Program, http://www.vdacs. virginia.gov/ plant&pest/hivegrant.shtml; Tennessee's Hive Grant, http://cookevillebeekeepers.com/tba-hive-grant; and Washington State beekeepers Association grants, http://wasba.org/about-wsba/wsba-grants/.

[104] EPA press release, "EPA Awards Almost Half a Million in Funding to Three Universities for Projects to Reduce Pesticide Risk Including Risks to Bees," January 8, 2014.

[105] T. Stecker, "New House Subcommittee Chief will Push Forward on Bee Issues," Environmental and Energy Policy, December 16, 2014. See also "Subcommittee Examines Research Efforts to Combat Pests and Diseases of Pollinators," House Subcommittee on Horticulture, Research, Biotechnology, and Foreign Agriculture hearing, April 29, 2014.

[106] For more information, see CRS Report R43076, The 2014 Farm Bill (P.L. 113-79): Summary and Side-by-Side. Many of these provisions were broadly extended through FY2013, as part of the one-year extension of the farm bill extension in the in the American Taxpayer Relief Act of 2012 (ATRA, P.L. 112-240).

[107] House-Passed 2013 farm bill, H.R. 2642, including text of H.R. 3102 (§12315).

[108] 7 U.S.C. §5925.

[109] 7 U.S.C. 5925(G) (High-Priority Research and Extension Initiatives; Pollinator Protection).

[110] For more information, see USDA, Forest Service, "Obtaining a Special-Use Authorization with the Forest Service: The Application Process," http://www.fs.fed.us/specialuses/ documents/broch.htm.

[111] USDA press release, "Agriculture Secretary Announces $3 Million for a New Program to Improve Pollinator Health," February 25, 2014. See also Honey Bee Pollinator Guidance, FY2014 (NB 300-14-14 LTP), January 2014.

[112] USDA, NRCS, "USDA to provide $4 million for honey bee habitat," October 29, 2014.

[113] Letter to Tom Vilsack, USDA Secretary, by numerous groups including the Pollinator Partnership, the American Beekeeping Federation, and the American Honey Producers Association, January 15, 2014. A "Varroa Mite Summit" is currently being organized by USDA.

[114] USDA, "Emergency Assistance for Livestock, Honeybees, and Farm-Raised Fish Program (ELAP)," October 2011. For more information, see CRS Report RS21212, Agricultural Disaster Assistance.

[115] USDA factsheet, "Emergency Assistance for Livestock, Honeybees and Farm-Raised Fish Program (ELAP), Honeybee Assistance," April 2014, http://www.fsa.usda.gov/Internet/FSA_File/elap_honeybee_fact_sht.pdf. Beekeepers can contact their local USDA/FSA office to apply (http://offices.sc.egov.usda.gov/locator/app?agency=fsa).

[116] FSA has established a normal mortality rate for honeybee colony losses of 17.5%.

[117] USDA, "USDA Disaster Assistance to Help Thousands of Honeybee, Livestock and Farm-Raised Fish Producers," press release #0258.14, November 24, 2014.

[118] 79 Federal Register 161: 49279-49280, August 20, 2014.

[119] R. Conniff, Green Highways: New Strategies to Manage Roadsides as Habitat," environment360, Yale University, June 10, 2013, http://e360.yale.edu/feature/green_highways_new_strategies_to_manage_roadsides_as_habitat/2661/.

[120] For information, see Pollinator Partnership/North American Pollinator Protection Campaign (NAPPC), "Support Highways Bettering the Economy and Environment Act (Highways BEE Act)," http://pollinator.org/BEEAct.htm.

[121] See, for example, press release on the bill by Representative Earl Blumenauer's office. For other related information, see CRS Report R42855, Bee Health: The Role of Pesticides.

In: Bee Health
Editor: Cristina Weaver
ISBN: 978-1-63482-671-6

Chapter 2

BEE HEALTH: THE ROLE OF PESTICIDES[*]

Renée Johnson and M. Lynne Corn

SUMMARY

Over the past few decades there has been heightened concern about the plight of honey bees as well as other bee species. Given the importance of honey bees and other bee species to food production, many have expressed concern about whether a "pollinator crisis" has been occurring in recent decades. Although honey bee colony losses due to bee pests, parasites, pathogens, and disease are not uncommon, there is the perception that bee health has been declining more rapidly than in prior years, both in the United States and globally. This situation gained increased attention in 2006 as some commercial beekeepers began reporting sharp declines in their honey bee colonies. Because of the severity and unusual circumstances of these colony declines, scientists named this phenomenon colony collapse disorder (CCD). Since then, honey bee colonies have continued to dwindle each year, for reasons not solely attributable to CCD. The U.S. Department of Agriculture (USDA) reports that CCD may not be the only or even the major cause of bee colony losses in recent years. In the United States, USDA estimates of overwinter colony losses from all causes have averaged nearly 30% annually since 2006.

The precise reasons for honey bee losses are not yet known. USDA and most scientists working on the subject seem to agree that no research

[*] This is an edited, reformatted and augmented version of a Congressional Research Service publication, No. R43900, dated February 17, 2015.

conclusively points to one single cause for the large number of honey bee deaths. This general conclusion was reconfirmed in a 2013 joint report by USDA and the U.S. Environmental Protection Agency (EPA). Reasons cited for bee declines include a wide range of possible factors thought to be negatively affecting pollinator species. However, one issue widely noted is the role that pesticides—in particular, neonicotinoid pesticides—might play in overall bee health. Pesticides are the focus of this report. Pesticides are among many identified factors known to affect bee health, including pests and diseases, diet and nutrition, genetics, habitat loss and other environmental stressors, and beekeeping management issues, as well as the possibility that bees are being negatively affected by cumulative, multiple exposures and/or the interactive effects of several of these factors.

The focus of this report on bee exposure to pesticides is not intended to imply that pesticides are any more important in influencing the health and wellness of bees than other identified factors influencing bee health. Pesticides are one of many influences on bee health. The current state of knowledge on pesticides and bee health is summarized in the USDA-EPA report:

> it is not clear, based on current research, whether pesticide exposure is a major factor associated with U.S. honey bee health declines in general, or specifically affects production of honey or delivery of pollination services. It is clear, however, that in some instances honey bee colonies can be severely harmed by exposure to high doses of insecticides when these compounds are used on crops, or via drift onto flowers in areas adjacent to crops that are attractive to bees.

Some experts emphasize research supporting the hypothesis that “total pesticide load” is an important influence on honey bee health, probably in combination with mite infestation, poor nutrition, viruses, and perhaps other stressors.

The past two farm bills (P.L. 110-246, P.L. 113-79) provided for increased funding for bee research, among other types of support to protect pollinators. Other bills in the 113th Congress addressed pesticide issues more directly. H.R. 2692 would have suspended registrations of neonicotinoids and banned new registrations of any pesticide in some cases. Another bill, H.R. 5447, would have amended U.S. pesticide laws to expedite the review and approval of products to control “parasitic pests” in managed commercial bee colonies, and would have required USDA and EPA to evaluate threats to pollinators and the availability of pesticides to manage bee pests.

There is increasing concern both in the United States and globally about whether a "pollinator crisis" has been occurring in recent decades. Reports worldwide indicate that populations of managed honey bees, wild bees, and native bees have been declining, with colony losses in some cases described as severe or unusual.[1] Other reports indicate that many insect pollinator species may be becoming rarer, which some say may be a sign of an overall global biodiversity decline.[2]

Many reasons are cited for bee population declines, including bee pests and diseases, diet and nutrition, genetics, habitat loss, agricultural pesticides, and beekeeping management. Because pesticides have been the focus of concerns in Europe and in the United States, this CRS report briefly describes recent scientific research and analysis regarding the potential role of pesticides among the factors affecting the health and well-being of bees. The report concludes with a summary of recent regulatory activity regarding neonicotinoids, a type of pesticide, and also provides the statutory authority and regulatory activities related to pesticide use at the U.S. Environmental Protection Agency (EPA), the federal agency charged with assessing risks and regulating U.S. sale and use of pesticides.

The focus of this report on bee exposure to pesticides is not intended to imply that pesticides are any more or less important in influencing the health and wellness of bees than any of the other identified factors influencing bee health.

BACKGROUND AND INTRODUCTION

In the United States, honey bee colony losses due to bee pests, parasites, pathogens, and disease are not uncommon. However, in late 2006, concerns about honey bees gained heightened attention when commercial beekeepers along the East Coast began reporting sharp declines in their bee colonies. Because of the severity and unusual circumstances of these colony declines, scientists named this phenomenon colony collapse disorder (CCD). This issue was legislatively active in the 110th Congress and resulted in increased funding for honey bee research, among other types of farm program support to protect pollinators, as part of the 2008 farm bill (Food, Conservation, and Energy Act of 2008, P.L. 110-246). The 2014 farm bill (P.L. 113-79) reauthorized and expanded upon many of these provisions.[3]

The U.S. Department of Agriculture (USDA) reports that U.S. beekeepers continue to lose colonies each year. Since 2006, USDA estimates that overwinter[4] bee colony losses have averaged nearly 30% annually.[5] However, USDA reports that in 2012/2013, "there were more colonies that dwindled away" rather than suffering from CCD, which is characterized by a sudden loss in bee colony populations and the absence of dead bees.[6] USDA also claims that "beekeepers did not report CCD as a major cause of colony loss" for overwinter losses reported in its 2012- 2013 and 2011-2012 surveys. These data are tracked for managed honey bees only.

Comparable data and information is not collected available for native or wild bee species. Such data collection is complicated by sheer number and solitary nature of native and wild bee species.

Managed Honey Bees and Wild Bee Species

Honey Bees

Honey bees (*Apis mellifera*; Family: Apidae) are the most well-known bee species. However, honey bees are only one of the world's estimated 17,000 described bee species, and one of the estimated total of 20,000 to 30,000 bee species worldwide. Honey bees are not native to North America, but were introduced by European settlers in the 1600s. Honey bees are considered to be "social" bees in that they have a single egg-laying queen and sterile worker bees that tend to work together in a highly structured social order, consisting of cooperation and division of labor within a colony, as well as the presence of two generations in a single nest at the same time. Social behavior allows bees to be domesticated and managed.

Some types of native bees may also be managed, such as bumble bees, orchard bees, and alfalfa leaf-cutting bees. Some of these bees (such as bumble bees and some types of stingless bees) exhibit some of the social behaviors commonly associated with honey bees. Some bumble bee species are managed in controlled environments to pollinate greenhouse tomatoes.

Wild Bee Species

An estimated 4,000 species of bees are native to North America. With few exceptions, most of these are wild and not managed. The five most common families of native bees in North America are Andrenidae, Apidae, Colletidae, Halictidae, and Megachilidae. Andrenid bees are all ground nesters, and mostly comprise a large family of dark, nondescript bees, although some are colorful.

> Bees in the large Apidae family include not only honey bees, but also bumble bees (such as *Bombus* spp.), carpenter bees, squash or gourd bees, and others. Most types of wild bees are "solitary" bees and do not have long-lived colonies. About 70% of native bee species are solitary ground nesting bees, and about 30% are solitary wood nesting bees. About 45 species of native bees in the United States are social bumble bees. Compared to most native species, bumble bees (*Bombus*) are better studied and, as noted by USDA, among the most effective crop pollinators.

To date, the precise reasons for honey bee colony losses are still unknown. USDA and most scientists working on the subject seem to agree that none of the research conclusively points to one single cause for the large-scale number of honey bee deaths. This general conclusion was reconfirmed in a May 2013 report by USDA and the U.S. Environmental Protection Agency (EPA), *National Stakeholders Conference on Honey Bee Health* (commonly referred to as the "USDA-EPA joint report").[7] A 2007 study by the National Research Council (NRC) of the National Academy of Sciences, *Status of Pollinators in North America* (referred to here as the 2007 NRC study), also provides a detailed scientific context for bee health. A series of other reports documenting the findings of USDA's ongoing research also describes the many factors affecting honey bees.[8] See **Figure 1**.

Reasons cited for bee population declines include a wide range of possible factors. Potential identified causes include bee pests and diseases, diet and nutrition, genetics, habitat loss and other environmental stressors, agricultural pesticides, and beekeeping management issues, as well as the possibility that bees are being harmed by cumulative, multiple exposures and/or the interactive effects of each of these factors.

One issue widely reported in the media is the potential role that pesticides—in particular, neonicotinoid pesticides—might play in overall bee health. As one of the potential causes of honey bee colony declines, this report addresses what role, if any, pesticides play in influencing the health and wellness of bees. Regarding honey bee health, the current state of knowledge of pesticides was summarized in a 2013 report by USDA and EPA:[9]

> There is broad consensus among all stakeholders that pesticide use should not affect honey bees in such a way that (1) honey production is reduced or (2) pollination services provided by bees are threatened.... However, it is not clear, based on current research, whether pesticide

exposure is a major factor associated with U.S. honey bee health declines in general, or specifically affects production of honey or delivery of pollination services. It is clear, however, that in some instances honey bee colonies can be severely harmed by exposure to high doses of insecticides when these compounds are used on crops, or via drift onto flowers in areas adjacent to crops that are attractive to bees.

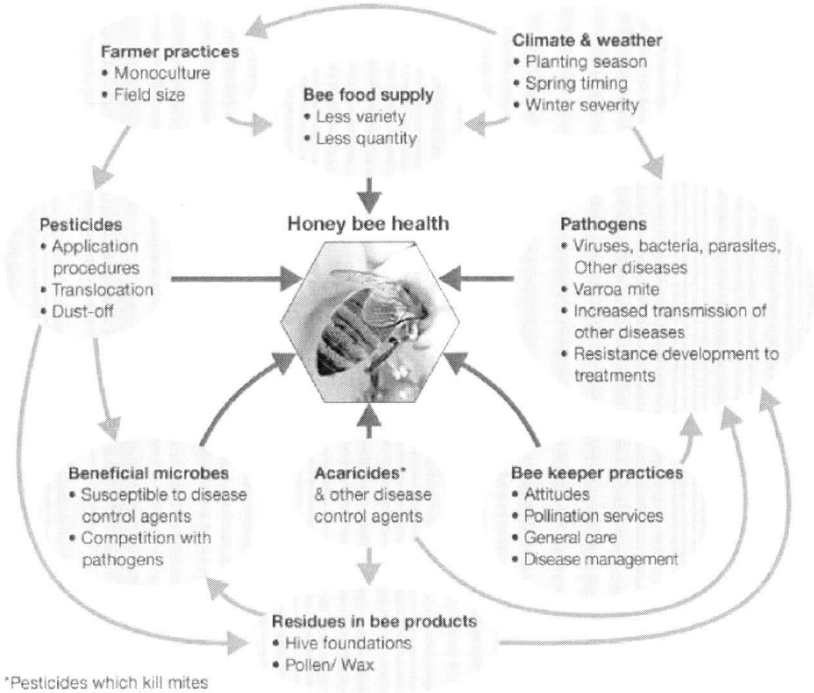

Source: OPERA Bee Health in Europe, 2013, http://www.operaresearch.eu/.

Figure 1. Stress Factors in Honey Bee Populations.

This report examines in greater detail the role of pesticides, providing a summary of selected scientific literature. The relative importance of pesticides in U.S. or global bee health is a subject of numerous research projects, some of which are discussed in this report.

Some groups have expressed concern about the assessment of most experts that the causes of pollinator health concerns are multifaceted and may involve the interaction of multiple factors, since this may deflect attention from the potential role of exposure to pesticides. On the other hand, some groups appear unwilling to acknowledge that pesticide exposure may play an important role in pollinator health concerns, especially if this acknowledgment leads to restrictions or reductions in the use of certain pesticides or related crop pest controls.

This report provides information regarding the potential role of pesticides in the health of bee colonies, and also the importance of pesticides relative to other influences on bee health. The report provides general information about the nature of pesticides, pesticide uses, and pesticide regulation in the United States, as well as more specific information about the registration status of a class of pesticides known as neonicotinoids, which have been implicated in some studies concerning honey bee colony declines. This report also describes a range of options to address pesticide exposure by bees, including implementing crop- and/or product-specific best management practices (BMPs) regarding pesticide use and applications. Some U.S. cities as well as some other countries, including Canada and those in Europe, have opted to institute restrictions on the use of certain pesticides. Congress has introduced similar legislation, but has also considered alternative policy options to address this issue.

U.S. Pesticide Laws and Regulation

Pesticide Laws and Statutory Framework

Pesticides are broadly defined in U.S. law as chemicals and other products used to kill, repel, or control pests.[10] Familiar examples include pesticides used to kill insects (insecticides) and weeds (herbicides) that can reduce the yield, and sometimes harm the quality, of agricultural crops, ornamental plants, forests, and pastures, or wooden structures (e.g., through termite damage). But the broad legal definition of "pesticide" also applies to products with less familiar "pesticidal uses." For example, substances are pesticides when used to control mites, mold, mildew, and other nuisance growths in hives or on equipment. The term also applies to disinfectants and sterilizing agents, animal repellents, rat poison, and many other substances. An estimated 18,000 pesticide products are currently in use in the United States.[11] Pesticides vary greatly in toxicity, persistence in the environment, and ability to bioaccumulate up the food chain, as well as in the range of plants and animals that are likely to be affected in the event of exposure. Some are nearly nontoxic to some species but exquisitely toxic to other species.

All pesticides are regulated by EPA under the authority of the Federal Insecticide, Fungicide, and Rodenticide Act (FIFRA),[12] and approximately 5,800 pesticide products used in food production also are regulated under the Federal Food, Drug, and Cosmetic Act (FFDCA).[13] FIFRA requires EPA to regulate the sale and use of pesticides in the United States through product registration and labeling so as to prevent unreasonable adverse effects on people and the environment, taking into account the costs and benefits of various pesticide uses. FIFRA prohibits the sale of any pesticide in the United States unless it is registered (licensed) and labeled to indicate approved uses and restrictions. It is a violation of the law to use a pesticide in a manner that is inconsistent with the label instructions. EPA registers each pesticide product for each approved use. For example, a product may be registered for use on bee hives to control mites or as a seed treatment for corn.[14] In addition, FIFRA requires EPA to reregister pesticides first registered prior to 1984 and to review all registered pesticides periodically on a 15-year cycle, based on new data that meet current regulatory and scientific standards.

For the 600 or more active ingredients in pesticide products that are registered for use in food production, Section 408 of the FFDCA authorizes EPA to establish maximum allowable residue levels (also known as "tolerances") to ensure that human exposure to the pesticide ingredients in food and animal feed will be "safe." A "safe" tolerance is defined in the law as a level at which there is "a reasonable certainty of no harm" from the exposure, even when considering total cumulative and aggregate pesticide exposure of children. Under the FFDCA, foods (or animal feeds) with a residue of a pesticide ingredient for which there is no tolerance established, or with a residue level exceeding an established tolerance limit, are declared "unsafe" and "adulterated"; such foods cannot be sold in interstate commerce or imported to the United States. Pesticides may not be registered under FIFRA for use on food crops unless tolerances (or exemptions) have been established under the FFDCA.[15]

Pesticide Registration Process

When pesticide manufacturers apply to register an active ingredient for a pesticide, a commercial pesticide product, or a new use of a pesticide registered under FIFRA Section 3, EPA requires them to submit scientific data on toxicity and behavior in the environment. In evaluating a pesticide registration application, EPA assesses a range of potential human health and environmental effects associated with use of the product. EPA's process of registering a pesticide comprises a scientific, legal, and administrative procedure involving the ingredients of the pesticide; the particular site or crop where it is to be used; the amount, frequency, and timing of its use; and storage and disposal practices.[16] EPA may require data from any combination of more than 100 different tests, depending on the potential toxicity of active and inert ingredients and degree of exposure.

To register a pesticide for use on food, EPA also requires applicants to determine the amount of residue that could remain on crops, as well as on (or in) food products (such as corn syrup), assuming that the pesticide product is applied according to the manufacturers' recommended rates and methods. Based on the data submitted, EPA determines whether and under what conditions a proposed pesticide use would present an unreasonable risk to human health or the environment, and, for a food or residential use, whether its use would be safe. Some features of pesticides that might affect registration decisions include the specificity of the pesticide for the targeted pest, its toxicity to people who apply it, its tendency to persist in the environment over time, and its ability to bioaccumulate in animals higher in the food chain.

EPA specifically takes into account unintended harm to bees and available information for other nontargeted insects in its registration decisions. EPA requires studies to determine acute (short-term) toxicity of a pesticide on individual bees when they come into body contact with pesticide residue. EPA also collects reports on bee-kill incidents. If a pesticide appears to be very toxic to bees, EPA may require long-term studies of its effects.

If the risk is determined to be unreasonable or unsafe, EPA attempts to mitigate the risk by adjusting requirements on the label (for example, requiring a buffer zone around lakes and streams or requiring personal protective equipment for pesticide handlers). If the risk remains unreasonable or unsafe, EPA will refuse to register the pesticide. If the risk is determined to be reasonable and safe, registration is granted, and the agency specifies the approved uses and conditions of use, including safe methods of pesticide storage and disposal, which the registrant must explain on the product label. EPA can and often does require specific application methods to be printed on the product label to minimize environmental damage. For example, the label sometimes requires that application of certain pesticides occur only when bees are not foraging, when there is little wind, or in a granular form or as a seed coating rather than aerially, in order to minimize spray drift off property. Pesticide registrations are reviewed at least once every 15 years to consider new scientific information and may be reviewed at any time in response to reports of adverse effects and possible unreasonable risks from use of particular pesticides.

Neonicotinoid Pesticides

Neonicotinoids are a relatively new major class of insecticides and among the fastest-growing class of insecticides in modern crop protection. Developed in the 1980s, some products such as imidacloprid were first introduced in the early to mid-1990s, but not widely marketed until the mid-2000s. Neonicotinoids are systemic pesticides that, regardless of application method (spray, drip irrigation, granular spreading, or seed coating), once taken into the plant, migrate into all parts, including flowers, pollen, and nectar.[17] Neonicotinoids are related to nicotine and were developed as an alternative to highly toxic (to humans) organophosphate insecticides such as methyl parathion.[18] Active ingredients of some of the most commonly applied neonicotinoids include imidacloprid, clothianidin, and thiamethoxam. Others include acetamiprid, dinotefuran, nitenpyram, and thiacloprid. (See text box below.)

Neonicotinoid Insecticides

In the United States, neonicotinoid pesticides are available for use in about 140 crops and garden/horticultural products, and provide potent and systemic action for crop protection, particularly from pests in soil and sap-sucking pests. They are used for seed treatment on most corn, soybeans, canola, sunflower, cereal grains, sugar beets, and potatoes. They are also used as foliar sprays on field and many fruit and vegetable crops (such as apples, cherries, peaches, oranges, berries, leafy greens, tomatoes, and potatoes) and are applied to cereal grains, rice, nuts, and wine grapes. They may also be injected into tree roots or stems and sprayed on tree bark to protect against pests. A single application can provide protection for several months or years and may remain with the plant as it grows. Since these types of pesticides can persist in soil or plants for long periods of time, this raises the potential that bee exposure will persist, even at low levels. Neonicotinoids reportedly accounted for almost 25% of the global pesticide market, and imidacloprid was the largest selling insecticide in the world in 2009, with sales exceeding $1 billion.

More recent information suggests that the total global market for neonicotinoid pesticides exceeded $2.6 billion in 2011.

Neonicotinoid Active Ingredients:

- Acetamiprid
- Clothianidin
- Dinotefuran
- Imidacloprid
- Nitenpyram
- Thiocloprid
- Thiamethoxam

Major Trade Names: Admire®, Acceleron®, Axcess®, Attendant®, Belay®, Cruiser®, Gaucho®, Nitro Shield®, Poncho®, and Trimax Pro®.

Selected Home and Garden Products: ALOFT®, ARENA®, CriterionTM, DIY Tree Care Products, Fertilome®, FlagshipTM, Green Light®, Safari®, Hi-Yield®, Knockout Ready-to-Use products, Mallet®, Marathon®, Maxide®, Meridian®, Merit®, Monterey Once A Year products, Ortho®, Safari products, Surrender®, TranstectTM, XytectTM, and Zylam®, and also several Bayer Advanced and Bayer Environmental Science products (including "12- Month"; "2-in-1"; "All-in-One"; "Complete Brand"; "Dual Action"; "Hunter"; "Lesco Bandit"; "Season-Long"; termite; and Bayer's fruit, citrus, and vegetable products).

Major Manufacturers: Bayer Advanced, Syngenta Crop Protection, LLC, and The Scotts Company, as well as ArborSystems, Arysta LifeScience, Control Solutions, Inc., Gro Tec, Inc., Gulfstream Home and Garden, Lawn and Garden Products, Inc., Nufarm Americas Inc., OHP, Inc., Rainbow Treecare Scientific Advancements, Valent U.S.A. Corporation, and Voluntary Purchasing Groups, Inc.

Sources: P. Jeschkel and R. Nauen, "Neonicotinoids—from Zero to Hero in Insecticide Chemistry," *Pest Management Science*, Vol. 64, Issue 11 (November 2008), pp. 1084–1098; M. Tomizawa and J. E. Casida, "Neonicotinoid Insecticide Toxicology: Mechanisms of Selective Action," *Annual Review of Pharmacology and Toxicology*, Vol. 45 (September 2004), pp. 247-268; P. Jeschke, R. Nauen, M. Schindler, et al., "Overview of the Status and Global Strategy for Neonicotinoids," *Journal of Agricultural* and Food Chemistry, v. 59 (2011), pp. 2897-2908; and B. Essler, "Are These Nicotine-Like Insecticides Killing Bees," *Modern Farmer*, November 12, 2014. Also Center for Food Safety (http://www.centerforfoodsafety.org /files/pesticide_list_final_59620.pdf), April 2013.

Neonicotinoids are generally considered to be reduced-risk compared to some other types of pesticides, and have low toxicity to mammals, birds, and fish compared to some other types of pesticides. However, as their use has increased, so have concerns about their potential harm to birds, earthworms, aquatic insects, and insect pollinators, including bees.[19] They comprise a class of active ingredients that have come under considerable scrutiny with respect to their potential effects on bee health. The attention is partly due to an incident of misuse (that is, use not in accord with the pesticide label) of one neonicotinoid, imidacloprid, in Germany[20] that resulted in a large bee kill, as well as widespread beekeeper concerns about use of another neonicotinoid, clothianidin, and its impact on bees in France.[21]

Neonicotinoids are insect neurotoxins that vary in strength of their effect on honey bees.[22] The scientific evidence to date indicates that although neonicotinoids are highly toxic to bees exposed to relatively high levels, individual pesticides in this class are not the only cause of declining bee health, and pyrethroid exposures may be more significant.[23] Although neonicotinoids have been a focus of scientific, public, and political interest, they have not been proven to be the primary cause of declines in bee health. Some experts, however, emphasize that research studies support the hypothesis that "total pesticide load" is an important influence on honey bee health, probably in combination with mite infestation, poor nutrition, viruses, and perhaps other stressors.[24]

Possible Role of Pesticides in Bee Health

Bees can be exposed to numerous different types of pesticides applied to field crops and other types of plants in areas where they forage or maintain their hive.[25] In addition, beekeepers may also use pesticides registered for the control of bacteria, fungi, mites, and other bee pests. These pesticides are applied within and in the vicinity of hives.[26] Besides the active ingredients, pesticide products include other ingredients, such as "inerts" or adjuvants that are intended to improve delivery of the active ingredient to the target pest.[27]

Others are used to increase the toxicity of a pesticide, for example, by inhibiting breakdown of a pesticide by insets.[28]

Studies have shown that bees are exposed to pesticides in many ways throughout the foraging period: from planter exhaust material produced during the planting of treated seed; from the soil of both planted and unplanted fields; in flowers growing near these fields; as well as applications in or near bee hives.[29] Bees also sometimes are exposed to pesticides accidentally, either when pesticides are misused or misapplied or when they are used according to label directions to control pests in areas frequented by bees—for example, alongside roads or rights of way for the control of weeds, trees, or other pests; on or near commercial farm crops; or on or near fields, lawns, and gardens to control fleas, ticks, weeds, grubs, mosquitos, or other adult insects.

Figure 2 illustrates some significant paths of bee exposure to pesticides applied as a spray or as a soil or seed treatment (systemic). If bees happen to fly through a newly treated field or dust clouds from planting of seeds coated with pesticide or are orally exposed to pesticide in food or water, and if exposure is high enough, bees may be sickened or die from pesticide exposure. With respect to the role of pesticides in honey bee health, "[t]he most pressing research questions lie in determining the true pesticide exposure that bees receive and the effect, if any, that pervasive exposure to multiple pesticides have [sic] on the health and productivity of whole honey bee colonies."[30]

Pesticides are reported to have adverse local impacts on honey bees and some native bees. Widespread use of herbicides reduces habitat available to bees;[31] many pesticides are known to be lethal to bees, given sufficient levels of exposure; and some reports of local bee kill incidents have been well documented. Effects on individual bees may be lethal or sublethal depending on dose and other conditions of exposure.[32] A summary of the types of sublethal effects reported in bees exposed to pesticides includes[33]

- decreased navigation, orientation, and communication abilities;
- altered foraging behavior and motor activity;
- short- and long-term memory loss;
- impaired learning behavior and sensory detection;
- compromised immune functioning;
- increased susceptibility to diseases and pests;
- reduced fecundity (fertility and reproduction); and
- impaired reproduction and development.

Although pesticides have been shown to damage bee health, it is unclear whether the level of harm is sufficient to attribute pesticides as the single or as the major cause of honey bee population declines.

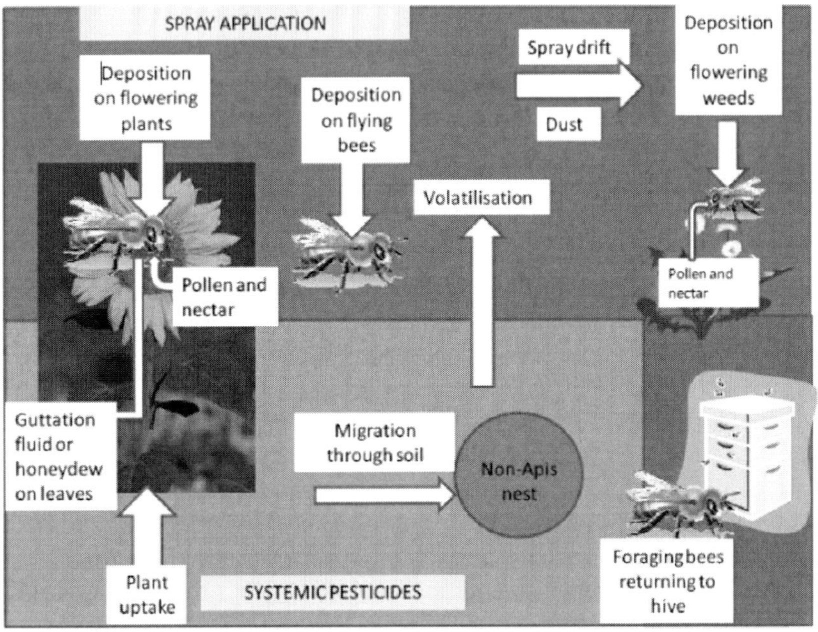

Source: European Food Safety Authority (EFSA), "Scientific Opinion on the Science Behind the Development of a Risk Assessment of Plant Protection Products on Bees (*Apis mellifera, Bombus* spp. and Solitary Bees)," Figure 3.1, EFSA-Q-2011-00417, May 2012, http://www.efsa.europa.eu/en/efsajournal/pub/2668.htm.

Figure 2. Major Routes of Exposure of Foraging Bees to Pesticides.

The **Appendix** provides a summary of selected scientific literature based largely on publications in peer-reviewed journals. In addition, a number of industry and advocacy groups have compiled literature reviews regarding pesticide effects on bees, not all of which are specifically discussed in this report.[34] In general, studies looking at impacts of pesticides on other animal species, such as birds, are also not addressed.[35]

BEST MANAGEMENT PRACTICES TO PROTECT POLLINATORS

Best management practices (BMPs) are available for beekeepers, crop producers, and pesticide applicators and include environmentally responsible pest management practices to reduce risk and minimize pesticide exposure in bees.[36] A number of states have also developed guidance to protect pollinators. Some of these resources are listed on the Pesticide Environmental Stewardship (PES) website[37] and on EPA's website.[38]

For beekeepers, guidance by apiculturists and university extension services emphasizes the importance of reducing the exposure of bees to insecticides with high toxicity, recommending that if such chemicals are used in an area where bees are foraging, steps may be necessary to reduce risk of poisoning. This involves both selecting a site for an apiary in an area with low pesticide risk and notifying "growers and applicators in the area, the county agent, and the State Apiary Inspector of the location of your hives.... If the insecticide to be used has a long residual life and is being applied to a plant where bees are foraging, it may be best to move your bees out of the area."[39]

For growers, recommendations involve reducing the hazards associated with insecticides, including avoiding the use of dusts, such as those from treated seeds, and using chemicals with reduced risk to bees whenever possible. Other recommendations include applying insecticides "in the late evening, night, or early morning when fewer bees will be foraging, and when spray drift and volatilization due to extreme heat are at a minimum"; not spraying "when winds favor drifting, and us[ing] ground applications instead of air where possible"; and avoiding "spraying when the crop or other plants in the field or nearby (including weeds) are in bloom."[40] Some agricultural groups provide such guidance to their growers. For example, the Almond Board of California recommends that growers avoid applying insecticides when plants are in bloom or when pollen is available and honey bees are feeding.[41] Additional grower BMPs are listed in the text box below.

Pesticide applicator BMPs recommend avoiding pesticide use when crops are blooming and applying pesticides to blooming crops only after bees are done foraging for the day and preferably at night.[42] Bayer Crop Science notes the importance of following pesticide label recommendations as "naturally beneficial to bees' safety," as well as cooperation between farmers and beekeepers "to optimize spray times and minimize exposure to foraging bees."[43]

Almond Board of California's "Top 10 Honey Bee BMPs"

1) Communication should occur between all pollination stakeholders about pest control decisions. These stakeholders can include beekeeper, bee broker, county agricultural commissioner, grower (owner/lessee), farm manager, pest control adviser (PCA), and pesticide applicator.
2) Agreements should include a pesticide plan that outlines which pest control materials may be used. The grower and the beekeeper should agree on which products may be applied if a treatment is deemed necessary. If deemed necessary, growers should give beekeepers 48-hour notice before treatment.
3) If applying pesticides, contact your local county agricultural commissioner and give advance notification to beekeepers with nearby managed hives.
4) Avoid applying insecticides during almond bloom until more is known, particularly about their impact on bee brood, or young developing bees in the hive. If treatment is necessary, only apply fungicides and avoid tank-mixing insecticides with fungicides.
5) Any fungicide application deemed necessary during bloom should occur in the late afternoon or evening, when bees and pollen are not present. This timing avoids contaminating pollen with spray materials.
6) Provide clean water for the bees to drink. This will ensure they spend more time pollinating the crop than searching for water. Cover or remove water sources before a pest control treatment, or empty and refill water after a treatment is made. Check water levels throughout bloom and refresh as necessary.
7) Do not directly spray hives with any pesticide spray application. Ensure that the spray-rig driver turns off nozzles when near hives. Spray applications that come in contact with bee hives could adversely affect bee health and the pollination of the crop.
8) Do not hit flying bees with any spray application materials. Bees that come in contact with agricultural sprays will not be able to fly because of the weight of spray droplets on their wings.
9) Report suspected pesticide-related bee incidents to the county agricultural commissioner's office. Bee health concerns cannot be addressed without the data from these incidents.

10) The beekeeper and the grower should agree on hive removal timing. The University of California recommends bee removal when 90% of the flowers on the latest blooming variety are at petal fall. Past this point, no pollination is taking place, and bees that forage outside the orchard (up to 4 miles) seeking alternate food sources and water will have a higher risk of coming in contact with crops that have been treated with an insecticide.

Source: Almond Board of California, "Honey Bee BMPs," http://www.almonds.com/newsletters/outlook/honey-beebmps-pest-management-and-honey-bees. See also http://www.growingproduce.com/nuts/top-10-honey-bee-bmps/.

Such BMP guidance is generally voluntary. Several studies suggest, in general, adoption of voluntary BMPs is characterized by both slow adoption rates (i.e., long lead times) and long lag times between BMP adoption and observed effects; also BMP implementation tends to be mostly self-funded.[44] Some claim therefore that reliance on voluntary agricultural BMPs regarding pesticide use and potential pollinator impacts is unlikely to produce timely behavioral changes and BMP adoption, even with a broad outreach and education program.[45]

In response to recent bee die-offs in Canada, federal agencies there have instituted additional protective requirements when using treated seed in corn and soybean production. These include required use of safer dust-reducing seed flow lubricants; adherence to safer seed planting practices; pesticide and seed package labels with enhanced warnings; and evaluation of the need for neonicotinoid treatment on certain commercial crops.[46] (More information on Canada's requirements is discussed in "Restrictions in Canada.")

Agro-chemical industry representatives maintain that unwanted pesticide exposure is best addressed through "effective product labeling and the implementation of meaningful stewardship actions that help minimize harmful interactions," along with "crop- and product-specific integrated pest management (IPM) practices and messaging to improve bee and pollinator safety."[47] However, some claim that because the use of systemic insecticides applied via seed coatings is mostly "prophylactic" (i.e., applied regardless of actual pest pressure) the use of such insecticides violates basic IPM principles, which recommend minimizing use of chemical pesticides through pest monitoring, maximizing the use of biological and cultural controls, applying chemical pesticides only when needed, and avoiding broad-spectrum, persistent compounds.[48] The text box below provides additional information on IPM practices.

Other guidance by apiculturists and university extension focuses on providing recommendations to private landowners and homeowners for proper use of pest control products on ornamental plants. One recommendation is to avoid applying any pesticides, including insecticides and fungicides, during bloom of ornamental plants that attract bees (e.g., heather, lavender, linden, rhododendron, and rose). It is also recommended that any pesticides be applied "only after flower petals have fallen, when ornamental plants are less attractive to bees," and that all specific requirements to protect bees on the pesticide label be strictly followed.[49] If pesticides are used when plants are in bloom, those that are less toxic to bees are recommended. Some recommend using certain pesticides—including products containing clothianidin, dinotefuran, imidacloprid, and thiamethoxam—only after flower petals have fallen, and avoiding soil drench or tree injection methods when using these products for plants known to attract bees because "these methods may contaminate nectar and pollen for up to several years after the insecticide is applied," or, alternatively, recommend against buying plants treated with insecticides containing these ingredients.[50]

What Is Integrated Pest Management (IPM)?

Integrated pest management (IPM) refers to practices that have been developed to improve pest control while also minimizing risks to beneficial species, including pollinators. IPM combines pest control with an understanding of the underlying ecology of the species and the environment where it occurs. The University of California Statewide Integrated Pest Management Program (UC IPM) defines IPM as:

> an ecosystem-based strategy that focuses on long-term prevention of pests or their damage through a combination of techniques such as biological control, habitat manipulation, modification of cultural practices, and use of resistant varieties. Pesticides are used only after monitoring indicates they are needed according to established guidelines, and treatments are made with the goal of removing only the target organism. Pest control materials are selected and applied in a manner that minimizes risks to human health, beneficial and nontarget organisms, and the environment.

EPA further states: "IPM relies on easy-to-implement, environmentally sensitive practices that prevent pests from becoming a threat. These practices involve monitoring and identifying pests and taking preventive action before pesticides are used. If pesticides are needed, methods such as targeted spraying may be used." In IPM strategies, pesticides may be considered part of IPM but generally are intended as a last resort, after other non-chemical methods have been exhausted or proved to be ineffective or are not available. According to UC IPM, pesticides are used "only when needed and in combination with other approaches for more effective, long-term control," and "are selected and applied in a way that minimizes their possible harm" to humans and other organisms, as well as the environment (air, soil, and water quality). The U.S. Fish and Wildlife Service (FWS) further states that in addition to reducing risks from pests and pest management-related strategies to people, property, resources, and the environment, one of the other benefits of IPM is the ability to "decrease or eliminate unnecessary pesticide use."

In January 2014, EPA announced it had awarded nearly $500,000 in agricultural grants for IPM practices to reduce the use of potentially harmful pesticides and lower risk to bees. These grants are intended to expand public-private stewardship efforts and reduce pesticide risk in agriculture. These IPM grants were awarded to Louisiana State University, Pennsylvania State University, and the University of Vermont.

Source: UC IPM website: http://www.ipm.ucdavis.edu; FWS, "Integrated Pest Management: Reducing the Risks to Pollinators from Pest Management Activities;" and EPA's press release, "EPA Awards Almost Half a Million in Funding to Three Universities for Projects to Reduce Pesticide Risk Including Risks to Bees," January 8, 2014. See also USDA, Natural Resources Conservation Service (NRCS), "Integrated Pest Management Plan Criteria Practice/Activity Code (114) (No.)," September 2011.

Consumer campaigns have been initiated to encourage businesses and home gardening centers to stop selling certain pesticides or plants treated with these products because of concerns about the effects on bees and other pollinators.[51] A consumer campaign initiated by the Center for Food Safety highlights that more than 60 commonly used home and garden products contain neonicotinoid pesticides and recommends that homeowners avoid certain commonly used pesticide products.[52] In addition, concerns have been raised about the use of mosquito control services by some homeowners, because of potential adverse effects to bees and other beneficial insects that might feed on plants or be exposed to pesticides within the sprayed areas, as well as effects to species on adjacent or nearby property due to drift.

Federal Agency Efforts

National Pollinator Health Strategy ("Presidential Memorandum")

In June 2014, the Obama Administration issued its Presidential Memorandum, "Creating a Federal Strategy to Promote the Health of Honey Bees and Other Pollinators," directing federal agencies to take steps to protect and restore domestic populations of pollinators.[53] It established a "Pollinator Health Task Force," co-chaired by USDA and EPA, with representatives of the Departments of State, Defense, the Interior, Housing and Urban Development, Transportation, Energy, and Education; among other agencies and offices.

The task force is directed to develop a National Pollinator Health Strategy, which is to include a Pollinator Research Action Plan to "focus federal efforts on understanding, preventing, and recovering from pollinator losses." Among the many activities expected to inform the action plan is "identification of existing and new methods and best practices to reduce pollinator exposure to pesticides, and new cost-effective ways to control bee pests and diseases." Task force member agencies will develop plans to increase and improve pollinator habitat. These plans may include "use of integrated vegetation and pest management," among other actions. Member agencies will also "make any necessary and appropriate changes to enhance pollinator habitat on federal lands through the use of integrated vegetation and pest management and pollinator-friendly best management practices." The task force is expected to release its National Pollinator Health Strategy in spring 2015.[54]

In November 2014, USDA and EPA held a number of public listening sessions to inform the task force members. Concerns about pesticide use were among the major discussion points. Commercial beekeepers, environmental groups, and some food businesses[55] continue to push for restrictions on pesticide use and question whether other factors, such as parasitic mites, are the primary driver behind bee declines. Most pesticide industry groups and commercial growers continue to encourage broader consideration beyond pesticides and claim that some alternative pesticides are more toxic than neonicotinoids.[56]

Currently, at EPA, the agency's pollinator strategic plan outlines efforts to advance scientific knowledge and assessment of pesticide risks to pollinators; improve management tools for mitigating risks to pollinators; and increase and broaden communication and public outreach.[57]

USDA-EPA Joint Report

In 2013, USDA and EPA published a joint report, *National Stakeholders Conference on Honey Bee Health.*[58] Among the goals of the conference were to "synthesize the current state of knowledge regarding CCD, bee pests, pathogens, and nutrition, potential pesticide effects on bees, and bee biology, genetics and breeding." The report's key findings include recommendations to address risks to honey bees from parasites and disease; increase the genetic diversity in bee colonies; and improve nutrition for honey bees. In addition, regarding pesticides, the report acknowledged the following needs:[59]

- **Collaboration and Information Sharing.** Best management practices (BMPs) associated with pesticide use and bees are known but are not widely or systematically followed by U.S. crop producers. "Informed and coordinated communication between growers and beekeepers" is needed, along with "effective collaboration between stakeholders on practices to protect bees from pesticides." Beekeepers have identified the need for "accurate and timely bee kill incident reporting, monitoring, and enforcement."
- **Additional Pesticide Research.** According to EPA, "[t]he most pressing pesticide research questions relate to determining actual pesticide exposures and effects of pesticides on bees in the field and the potential for impacts on bee health and productivity of whole honey bee colonies."

The National Honey Bee Health Stakeholder Conference Steering Committee is made up of representatives from Pennsylvania State University; from USDA's Office of Pest Management Policy (OPMP), National Institute of Food and Agriculture (NIFA), Agricultural Research Service (ARS), Animal and Plant Health Inspection Service (APHIS), Natural Resources Conservation Service (NRCS), and National Agricultural Statistics Service

(NASS); and from EPA's Office of Pesticide Programs (OPP). Much of the current research on bee health is being conducted by scientists at USDA and its Beltsville bee laboratory, by the USDA-supported Bee Informed Partnership, and by scientists at many of the land-grant universities nationwide.[60]

Environmental Protection Agency

EPA's Pesticide Registration Review of Neonicotinoid Pesticides

Neonicotinoid pesticide registrations are being reviewed by EPA. According to EPA:[61]

> Some uncertainties have been identified since their initial registration regarding the potential environmental fate and effects of neonicotinoid pesticides, particularly as they relate to pollinators. Data suggest that neonicotinic residues can accumulate in pollen and nectar of treated plants and may represent a potential exposure to pollinators.
>
> Adverse effects data as well as beekill incidents have been reported, highlighting the potential direct and/or indirect effects of neonicotinic pesticides. Therefore, among other refinements to ecological risk assessment during registration review, we will consider potential effects of the neonicotinoids to honeybees and other pollinating insects.

Review of several neonicotinoid pesticide registrations began December 2011, with review of imidacloprid starting a few years earlier, in December 2008 (**Table 1**). EPA aims to review all neonicotinoids as a group.

In July 2014, the Natural Resources Defense Council (NRDC) petitioned EPA to conduct an emergency review of the impacts of neonicotinoid pesticides on bees and to complete its review within one year.[62] At an October 2014 meeting, an EPA official indicated that the agency would move up its dates for its review of neonicotinoid pesticides from 2018-2019 to 2016-2017.[63]

Currently, among the neonicotinoid pesticides under review, some products are being used under a "conditional" registration, while some products are being used under an "unconditional" registration.

Table 1. Schedule for EPA Review of Neonicotinoid Pesticides

Chemical Name and Docket Number	Initiation	Data Generation	Projected Completion
Imidacloprid (EPA-HQ-OPP-2008-0844)	Dec. 2008	2010-2015	2016-2017
Clothianidin (EPA-HQ-OPP-2100-0865)	Dec. 2011	2013-2016	2017-2018
Thiamethoxam (EPA-HQ-OPP-2011-0581)	Dec. 2011	2013-2016	2017-2018
Dinotefuran (EPA-HQ-OPP-2011-0920)	Dec. 2011	2013-2016	2017-2018
Acetamiprid (EPA-HQ-OPP-2012-0329)	Dec. 2012	2014-2017	2018-2019
Thiacloprid (EPA-HQ-OPP-2012-0218)	Dec. 2012	2014-2017	2018-2019

Source: EPA's website, http://www2.epa.gov/pollinator-protection/schedule-review-neonicotinoid-pesticides.

FIFRA provides that EPA register a pesticide if, among other findings, it meets the statutory standard, namely: (1) the pesticide will perform its intended function without unreasonable adverse effects on human health and the environment, and (2) the pesticide will not generally cause unreasonable adverse effects on human health and the environment when used in accordance with widespread and commonly recognized practice. Under an unconditional registration, after reviewing the data and information submitted to support an applicant/registrant's pesticide product application under consideration, if EPA determines that the pesticide meets the statutory standard and there are no outstanding data requirements, the agency may approve an "unconditional" registration.[64] If, however, EPA finds that the pesticide meets the standard for registration, but there are outstanding data requirements, the agency may, under certain circumstances, grant a "conditional" registration.[65] Before granting a conditional registration, EPA must determine that, although an application lacks some of the necessary data, use of the pesticide would not significantly increase the risk of unreasonable adverse effects on the environment during the time needed to generate the necessary data. A product's "conditional" or "unconditional" status is determined on a product registration by product registration basis, and not for the pesticide's active ingredient as a whole.[66] An indication of a product's current status can be found at EPA's searchable label database.[67]

The U.S. Government Accountability Office (GAO) has conducted studies that are critical of EPA's pesticide registration process, particularly regarding the program's conditional registrations.[68] Environmental and other groups argue that conditional registrations are a loophole in EPA's requirements, allowing pesticides onto the market that might not otherwise be allowed under a thorough agency review.[69] These groups further claim that EPA has overused conditional registrations and that the agency does not have a reliable data-gathering system to track conditional registrations.[70] GAO's 2013 report states that "the total number of conditional registrations granted is unclear" due to database inaccuracies and confusion in EPA's recordkeeping system for tracking pesticides.[71]

In the process of reviewing registrations for neonicotinoids, EPA has revised its risk assessment process "to reflect advancements in the state of the science that underlie bee exposure and effects assessments."[72] The draft risk assessment policy was released to the public, and materials were distributed and discussed at a FIFRA Scientific Advisory Panel (SAP) meeting in September 2012.[73] The final risk assessment guidance was released in June 2014.[74] The guidance is founded on EPA's ecological risk assessment framework, integrates an analysis of honey bees and other pollinators within the agency's broader components of ecological risk assessment, and was prepared in collaboration with Health Canada's Pesticide Management Regulatory Authority and California's Department of Pesticide Regulation. The guidance describes the basic framework of the risk assessment process and the data used to support risk management decisions, and includes a process to assess both foliar spray applications and soil/seed treatment applications.

Some groups argue that EPA does not take into account chronic, sub-lethal effects in its risk assessments, nor synergistic effects with other compounds used in real field settings, such as fungicides, adjuvants, and inerts.[75] These groups also claim native and wild bee species are generally not addressed as part of EPA's risk assessment. Since most native bee species (about 70%) are ground nesting,[76] these groups further claim that the use of systemic insecticides poses risks to native bees from exposure through their habitats in the ground. Native and wild bee species are also generally not addressed as part of EPA's pesticide labeling approach (discussed in "EPA's New "Bee Advisory" Labeling Requirements").

As part of EPA's overall review, the agency has conducted a study of the benefits of neonicotinoid seed treatments for insect control in U.S. soybean production.[77] EPA's analysis concluded: "these seed treatments provide little or no overall benefits to soybean production in most situations. Published data indicate that in most cases there is no difference in soybean yield when soybean seed was treated with neonicotinoids versus not receiving any insect control treatment."[78] Reportedly, EPA also studied the potential benefits of neonicotinoid seed treatments to corn production but has not published those findings.[79] In 2013, EPA and USDA conducted a summit with stakeholders on reducing exposure to dust from treated seed and potential acute exposure of honey bees and pollinators to pesticides. Information and stakeholder presentations from the summit are at EPA's website.[80]

EPA's Consideration to Modify Thiamethoxam Tolerance Levels

In September 2014, EPA announced that it had received "several initial filings of pesticide petitions requesting the establishment or modification of regulations for residues of pesticide chemicals in or on various commodities," including a petition by Syngenta that EPA increase the allowable threshold for residues of thiamethoxam.[81] Syngenta's petition would apply to alfalfa, barley, corn and wheat, both the crop itself and the straw and stover[82] left over after cultivation. According to some reports, the petition seeks to increase the tolerance levels ranging from about 1.5 times current levels for stover from sweet corn to about 400 times current levels for hay from wheat.[83] EPA's review of this request is still pending. Thiamethoxam is one of the neonicotinoid pesticides banned for use in the European Union.

EPA's New "Bee Advisory" Labeling Requirements

In August 2013, EPA announced that it had developed new pesticide labels that prohibit use of some neonicotinoid pesticide products where bees are present. The new requirements apply to foliar applications of products containing imidacloprid, dinotefuran, clothianidin, thiamethoxam, tolfenpyrad, and cyantraniliprole.[84] Exceptions apply under certain conditions for agricultural crops and commercially grown ornamental plants. The new labels require a "Pollinator Protection Box" (or "bee advisory," **Figure 3**) and bee icon with information on routes of exposure and spray drift precautions, as well as new language added under "Directions for Use."

The bee advisory will, among other things, alert pesticide applicators to restrictions regarding certain pesticides when bees are present; clarify that pesticides cannot be applied until all petals have fallen; and emphasize the importance of avoiding pesticide drift (for example, due to wind) to other areas where bees may be present. The new bee icon will signal the pesticide's potential hazard to bees, and warns that direct contact and ingestion can harm pollinators. Information is at EPA's website.[85]

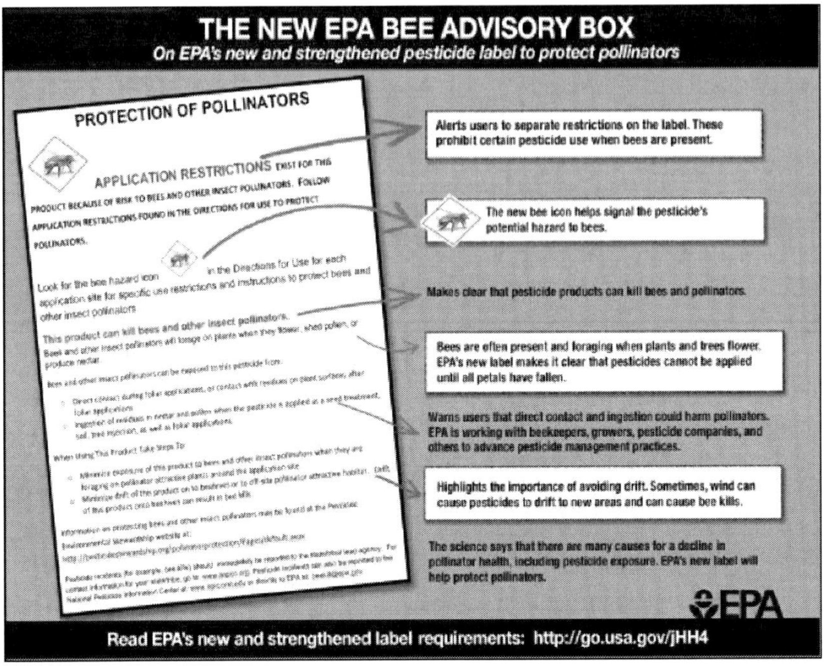

Source: EPA, http://www2.epa.gov/sites/production/files/2013-11/documents/bee-label-info-graphic.pdf.

Figure 3. EPA's Bee Advisory Box.

EPA's new labeling requirements have received mixed reviews from beekeepers, who argue that the new labels are inadequate and include exceptions that may make them less protective for bees and other pollinators.[86] Groups, such as the Pollinator Stewardship Council, have expressed concerns about EPA's label as well as concerns about the conditions under pesticide applications would be allowed. **Figure 4** shows some of this group's concern with EPA's label, including concerns that none of the terms in the label is defined and many are ambiguous (which they claim makes the requirements unenforceable), as well as the concern that the label refers applicators to a pesticide industry website, among several other issues.[87]

Some groups, including the Pollinator Stewardship Council, have voiced concerns about EPA's exceptions to the new labeling requirements, which would allow application under certain conditions.[88] Exceptions to the labeling requirements are as follows.

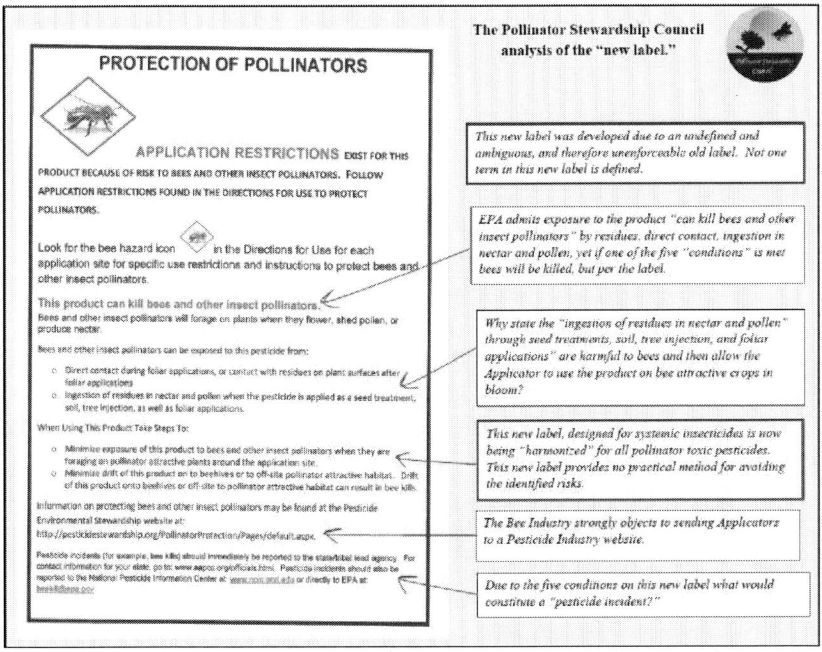

Source: Pollinator Stewardship Council. Other comments regarding exceptions to the label's requirements are at http://pollinatorstewardship.org/wp-content/uploads/2014/02/PSC-label-layout-of-concerns.pdf.

Figure 4. Pollinator Stewardship Council's Analysis of EPA's Bee Advisory Box.

For "**Crops Grown Under Contracted Pollination Services**," the label for some neonicotinoid pesticide products states: "Do not apply this product while bees are foraging" or "until flowering is complete and all petals have fallen." Exceptions are allowed for in cases where "an application must be made when managed bees are at the treatment site, the beekeeper providing the pollination services must be notified no less than 48 hours prior to the time of the planned application so that the bees can be removed, covered or otherwise protected prior to spraying."

For "**Food Crops and Commercially Grown Ornamentals not Under Contract for Pollination Services But Attractive to Pollinators**," the label for some products also states: "Do not apply this product while bees are foraging" or "until flowering is complete and all petals have fallen." Exceptions are allowed for in cases where:

- The application is made to the target site after sunset.
- The application is made to the target site when temperatures are below 55◦F.
- The application is made in accordance with a government-initiated public health response.
- The application is made in accordance with an active state-administered apiary registry program where beekeepers are notified no less than 48 hours prior to the time of the planned application so that the bees can be removed, covered, or otherwise protected prior to spraying.
- The application is made due to an imminent threat of significant crop loss, and a documented determination consistent with an IPM plan or predetermined economic threshold is met. Every effort should be made to notify beekeepers no less than 48 hours prior to the time of the planned application so that the bees can be removed, covered, or otherwise protected prior to spraying.

Among the expressed concerns regarding EPA's exceptions to its label advisory are:[89]

- harm caused by foliar application of the affected pesticides will be the responsibility of the beekeeper, including damage or die-off from applications allowed for under the labels exceptions;
- uncertainty about what constitutes notifying a beekeeper to move their bees, and concern that native bees will be harmed since they are not similarly managed;
- foliar application of adjacent or nearby sites may affect bees since they have a 3- mile to 7-mile forage range;
- exceptions to applications made when temperatures are below 55°F may affect bees that forage at temperatures as low as 45°F; and
- other labeling issues, including who decides when foliar treatments are needed, what are the criteria for determining whether treatment are needed, where should beekeepers transport bees during applications, and who determines what mitigation measures are appropriate.

Exceptions do not apply to non-agricultural crops, such as ornamental crops grown by homeowners, and the pesticide label requires that the product not be applied while bees are foraging or when plants are flowering.

EPA Support of State Pollinator Protection Plans

EPA is working with states to draft guidance on the development of EPA-approved, state-managed pollinator protection plans. Guidance is being developed by EPA's State FIFRA Research and Evaluation Group (SFIREG), and will outline a process for reviewing and accepting state pollinator protection plans as part of EPA's broader pollinator protection efforts related to pesticides. The draft guidance was presented at a December 2014 meeting of the Association of American Pesticide Control Officials (AAPCO),[90] and contained information on the core elements for EPA-approved state pollinator plans, the process for EPA review and updating of state plans, how to reference state plans through labeling requirements, and how to access recommended best management practices.[91] Pollinator protection plans are in place in several states, including California, Colorado, Florida, Mississippi, and North Dakota.[92] Other states are considering similar initiatives.

EPA Development of Voluntary Controls for Pesticide Drift

In October 2014, EPA announced a new voluntary Drift Reduction Technology (DRT) program to encourage the use of verified, safer pesticide spray products to reduce exposure and pesticide movement, and also to reduce costs to farmers from pesticide loss. EPA defines pesticide spray drift as the "movement of pesticide dust or droplets through the air at the time of application or soon after, to any site other than the area intended."[93] More information is at EPA's website.[94] A similar effort—the DriftWatch Specialty Crop Site Registry—provides for a voluntary communication tool between crop producers, beekeepers, and pesticide applicators, and is intended to protect high-value, pesticide-sensitive crops and commodities before applying pesticides.[95] The program is currently in use in many midwestern and mid-Atlantic states.

Fish and Wildlife Service Memorandum

In July 2014, the U.S. Fish and Wildlife Service (FWS) in the Department of the Interior announced that it would phase out the feeding of genetically engineered crops to wildlife and the use of neonicotinoid pesticides in all of its wildlife refuges by January 2016.[96]

Initially this decision was meant to phase out the use of neonicotinoids in Region 1 only, which covers Oregon, Washington, Idaho, Hawaii, and the Pacific Islands.[97] Under the guidelines, refuge managers will also need to comply with new mandatory requirements for all chemically treated seeds on refuge lands.[98]

In early 2013, Representative Austin Scott, chairman of a House Agriculture subcommittee, indicated that the committee would direct FWS to undergo "a uniform risk assessment process ... when making decisions on products already approved" under FIFRA and/or the Plant Protection Act.[99] To date, no such action has been untaken.

OTHER ONGOING EFFORTS

Restrictions in Some U.S. Cities and Counties

A number of cities and counties, among other local jurisdictions, have instituted restrictions on the use of neonicotinoid pesticides in an effort to protect pollinators. In March 2014, Eugene, Oregon, became the first U.S. city to restrict neonicotinoid use. (Despite the ban, a mass die-off of bees was reported in June 2014 at an apartment complex in northwest Eugene. It was attributed to insecticide sprayed on blooming linden trees.) Since then other cities have also instituted bans, including Seattle and Spokane in Washington; Shoreline, Minnesota; and Skagway, Alaska. Other cities and counties, such as Tucson and Pima County in Arizona, and Boulder County in Colorado, also are considering restrictions. Some state legislatures, including Alaska, California, Maryland, Minnesota, New York, New Jersey, Oregon, and Vermont, have considered legislation to further study and/or restrict neonicotinoid use. Some localities also have instituted other types of pesticide restrictions, such as with disclosure laws in the county of Kaua`i in Hawaii. Other localities have implemented programs to encourage households and municipalities to pledge to create "pollinator-friendly, toxin-free ecosystem habitats," such as in the city and county of Denver, Colorado. Some state agencies are actively studying the issue.[100]

In addition, several states—California, Colorado, Florida, Mississippi, and North Dakota—have state pollinator protection plans in place, and other states are considering similar initiatives.[101] (For information, see "EPA Support of State Pollinator Protection Plans")

Restrictions in the European Union

Like the rest of the world, the countries of the EU have experienced overwinter honey bee colony losses. As reported by the nonprofit honey bee research association, COLOSS,[102] overwinter colony losses averaged 9% in 2013/2014, the lowest since the group began collecting such data in 2007.[103] Across countries, losses ranged from 6% to 12%. This is lower than results reported for 2012/2013, when overwinter losses ranged from 6% to a high of 37% across all reporting countries.[104] Another report commissioned by the European Commission and EU member states concluded honey bee colony mortalities were "better than previously expected" and "higher than normal in certain countries, with significant regional (and possibly temporal) differences."[105]

In response to concerns about declining bee populations, in May 2013, the European Commission (EC) adopted Regulation No 485/2013 banning the use of certain neonicotinoid pesticides for a period of two years, among other proposed limits on the use of other pesticides.[106] The regulation includes the following provisions:

- use of three neonicotinoid pesticides—clothianidin, imidacloprid, and thiamethoxam—is restricted for seed treatment, soil application (granules), and foliar treatment on bee attractive plants and cereals;
- any authorized uses are available to professionals only;
- any exceptions to these restrictions are limited to treating bee-attractive crops in greenhouses or in open-air fields after flowering; and
- restrictions are to be maintained for two years, effective December 1, 2013.

As new information becomes available, the EC will review the approval conditions for these three pesticides and take into account relevant scientific and technical developments. Previous mitigation measures (EC Regulation No 1107/2009) restricted spraying of insecticides on bee-attractive flowering crops.[107]

The EC's 2013 regulatory action was in response to the European Food Safety Authority's (EFSA)[108] scientific assessment that identified "high risks for bees for some uses of three neonicotinoids (imidacloprid, clothianidin and thiamethoxam) and fipronil," such that "the approval criteria of these pesticides were no longer satisfied."[109] Regarding the three restricted pesticides, EFSA identified "high acute risks" for bees from exposure to pesticides associated with the production of several crops such as corn, grains, and sunflower, as well as exposure to residue in pollen and nectar in certain crops.[110]

The EU Commission has further proposed to restrict the use of fipronil, an insecticide which EFSA has identified as posing an acute risk to Europe's honey bees when they are exposed to seeds treated with the chemical.[111]

Other reports, such as a study published by the United Kingdom's Department for Environment, Food and Rural Affairs (DEFRA), dispute some of the findings linking bee health and exposure to neonicotinoids.[112] Other studies also highlight the likely multifaceted nature of possible factors contributing to pollinator declines.[113] Researchers in Europe continue to study this issue and reportedly are in the process of completing a large-scale study reviewing the landscape-scale effects of neonicotinoid use on crops.[114] Some member state officials are also questioning the ban and its possible effects on agricultural production.[115]

Restrictions in Canada

In Canada, the Canadian Association of Professional Apiculturists (CAPA) reports that overwinter honey bee colony losses in 2013/2014 averaged 25% across Canada, but some provinces, such as Ontario, reported wintering losses of 58%.[116] In addition, in 2012, Health Canada's Pest Management Regulatory Agency (PMRA) began reporting higher bee mortalities associated with pesticide applications in some areas as part of its pesticide incidents reporting.[117]

In April 2012, PMRA reported a "significant number of honey bee mortality reports" in the provinces of Alberta, Manitoba, Saskatchewan, Nova Scotia, Quebec, and Ontario. Most reports were from southern Ontario, involving more than 40 beekeepers and 240 different locations, with also one report from Quebec involving eight bee yards. Health Canada concluded "an unusually high number of reports of honey bee mortalities were received from beekeepers in corn growing regions of Ontario and Quebec.... Timing and location of these honey bee mortalities appeared to coincide with planting corn seed treated with insecticides."[118]

PMRA's preliminary findings conclude "there is an indication that pesticides used on treated corn seeds may have contributed to at least some of the 2012 spring bee losses that occurred in Ontario."[119] The agency continued to "receive a significant number of pollinator mortality reports from both corn and soybean growing regions of Ontario and Quebec, as well as Manitoba" in spring 2013.[120] A 2013 field study by researchers at the University of Guelph reported no link between bee health and pesticide applications.[121]

Following this investigation, PMRA determined that "current agricultural practices related to the use of neonicotinoid treated corn and soybean seed are affecting the environment due to their impacts on bees and other pollinators."[122] The agency implemented measures to reduce honey bee exposure to dust generated during planting of treated corn seed, including "communication of best practices to reduce the exposure of honey bees, labelling of treated seed, a treated seed dust standard, and development of technical solutions to reduce dust, including developments in the areas of seed coating quality, seed flow lubricants, planting equipment, and disposal of treated seed bags."[123] For the 2014 planting season, PMRA issued additional protective measures for corn and soybean production.[124] PMRA has continued to re-evaluate nitro-guanidine neonicotinoids to determine if further regulatory action was needed.[125] PMRA is reportedly also planning to implement label changes similar to those being considered by EPA in the United States (see "EPA's New "Bee Advisory" Labeling Requirements").

In November 2014, the province of Ontario announced a proposal to reduce the use of neonicotinoid-treated corn and soybean seed (based on acreage of treated seeds planted) by 80%.[126] If approved, the new regulations would go into effect by July 1, 2015, prior to the 2016 planting season. Manufacturers of the chemicals claim the decision is not supported by science. The Canadian industry association, CropLife Canada, is also urging the Canadian government to broaden its examination of bee population declines beyond pesticides and neonicotinoids.[127] The Canadian government continues to study this issue.[128]

In September 2014, two large beekeeping operations filed a class-action lawsuit[129] on behalf of Canadian beekeepers against several chemical manufacturers, claiming thiamethoxam (and its predecessor, imidacloprid) and its breakdown product clothianidin led to more than C$450 million (about US$410 million) in total damages from 2006 to 2013.[130] The alleged chronic effects of the use of these pesticides include "bee deaths; impaired reproduction; immune suppression; behavioral abnormalities resulting in hive loss; reduced honey production; impacts on the quality of honey; contamination of hive equipment; loss of queen bees; breeding stock; and difficulties fulfilling honey product or pollination contracts." The lawsuit alleges that Bayer and Syngenta were "negligent in their design and development of the neonicotinoid pesticides," and were and continue to be "negligent in their distribution and sale of the neonicotinoid pesticides," as well as "negligent in permitting or failing to prevent the damages caused by the neonicotinoids to the beekeepers." The plaintiffs further claim that Bayer and Syngenta "knew or ought to have known that the neonicotinoids would cause damage to the property" of beekeepers, since they allege that the harm to the beekeepers was "reasonably foreseeable."

LEGAL AND OTHER ACTIONS BY SOME INTEREST GROUPS

Selected Actions by Environmental Groups

A number of environmental and food safety advocacy groups, along with individual beekeepers, have remained active in pressing federal and state authorities on policies and issues pertaining to the approval and use of neonicotinoid pesticides, among other agrochemicals. Among these are the Center for Food Safety (CFS); Earthjustice; Pesticide Action Network North America (PANNA); Beyond Pesticides; Friends of the Earth; Sierra Club; and the Center for Environmental Health. These groups have published a series of widely available reports supporting their policy positions and/or have posted literature reviews on their organization's websites, and also have undertaken a series of legal challenges in these matters.[131]

In March 2013, CFS and a coalition of four beekeepers and five environmental and consumer groups filed a lawsuit against the EPA to stop the use of pesticides containing clothianidin and thiamethoxam, which beekeepers in the case claim is damaging the central nervous system of their bees.[132] The lawsuit also challenges EPA's risk assessment framework for determining whether pesticides harm pollinators and questions EPA's approval of certain neonicotinoids. An opening brief was filed in December 2013 further challenging EPA's risk assessment framework and requesting that the court reverse EPA's decision to register sulfoxaflor.[133] In July 2014, CFS, Earthjustice, PANNA, and Beyond Pesticides challenged California's approval of new agricultural uses for two neonicotinoid pesticides, Venom Insecticide and Dinotefuran 20SG.[134]

Previously, in 2005, two beekeepers residing in Minnesota sued the state's Department of Natural Resources for losses incurred from spraying of a pesticide, carbaryl. In that case, the pesticide users reportedly offered to settle, and the state's Department of Natural Resources stopped using the pesticide.[135]

These legal challenges follow previous requests for EPA to restrict the use of certain pesticides. In a March 2012 citizen petition, CFS and a coalition of beekeepers along with PANNA filed an "Emergency Petition" with EPA asking for a suspension of the use of clothianidin until it is proven safe to pollinators and the environment.[136] EPA responded in part to the petition in July 2012.[137] A few weeks later, EPA announced that it was denying the request to suspend registrations "to prevent imminent harm" because the petitioners did not meet the burden of proof for registration suspension.[138] However, EPA received public comments on this decision and may revisit its decision as it reviews neonicotinoid registrations. Again, in July 2014, NRDC petitioned EPA to conduct an emergency review of the impacts of neonicotinoid pesticides on bees, and to complete its review within one year.[139] For more information, see "EPA's Pesticide Registration Review of Neonicotinoid Pesticides."

In June 2013, a coalition of advocacy groups, including CFS, sent a letter to President Obama asking him to order EPA to ban use of neonicotinoid insecticides for at least two years.[140] The letter asserts that the registration process lacks consideration of pesticide effects on colonies and focuses exclusively on acute mortality, rather than on sublethal effects of repeated exposure. Another letter followed in January 2015 from commercial beekeepers, environmental groups, and some food businesses, who continue to push for restrictions on pesticide use.[141]

Selected Actions by Agrochemical Companies

Some major agrochemical companies—in particular, Syngenta AG (a global Swiss agribusiness company) and Bayer AG (a global German company)—have been actively engaged in defending the use of neonicotinoid pesticides, among other agrochemicals, and claim there is no evidence to support a systematic correlation between honey bee colony mortality and the use of neonicotinoids.[142] Bayer's bee care pamphlet claims "poor bee health is correlated with the presence of *Varroa*, a parasitic mite, viruses and many other factors, but not with the use of insecticides."[143] These companies and other industry groups, as well as the trade association CropLife Amercia,[144] have published commissioned reports supporting their policy positions,[145] and have supported various pollinator health[146] as well as lobbying efforts.[147]

According to news reports, both Syngenta and Bayer are legally challenging the EU's neonicotinoid ban.[148] Syngenta claims these pesticides are not responsible for the pollinator population declines, which they claim are instead the result of disease, viruses, habitat loss, and poor nutrition.[149] According to reports, Syngenta wants to reverse the EU ban but also wants to be compensated for damages and "defend our reputation which has been significantly damaged."[150] Syngenta claims the EU's decision to ban the pesticides is based on "a flawed process, an inaccurate and incomplete assessment by the European Food Safety Authority and without the full support of EU Member States."[151]

Syngenta also has petitioned EPA to increase the allowable threshold for residues of thiamethoxam. EPA published its proposal in September 2014.[152]

CONGRESSIONAL CONSIDERATION

Given continued concerns about the health and well-being of honey bees and other pollinators, this issue has continued to be legislatively active. The 2014 farm bill (P.L. 113-79) reauthorized and expanded provisions supporting research on honey bees and other pollinators that were enacted in the 2008 farm bill. In previous Congresses, bills were introduced to promote and improve habitat for honey bees and wild bees, among other pollinators.[153] Still other introduced legislation would address a range of pesticide issues affecting pollinators. Over the past few years, Congress has conducted hearings on this issue[154] and hosted a series of congressional briefings to discuss these and related issues representing a broad range of interested groups.[155]

For the 114th Congress, Representative Rodney Davis, the incoming chairman of the House Agriculture Subcommittee on Biotechnology, Horticulture, and Research, is quoted as saying that bee health will be a top issue for his subcommittee.[156]

Saving America's Pollinators Act

In the 113th Congress, Representatives Earl Blumenauer and John Conyers Jr. introduced H.R. 2692, the Saving America's Pollinators Act of 2013. The bill would have suspended registrations of neonicotinoids and banned new registrations of any pesticide for use on "bee attractive plants, trees, and cereals." This bill was initially introduced, in part, as a response to reports that

50,000 bees were found dead in a suburban shopping-center parking lot in Wilsonville, Oregon, in June 2013, reportedly due to exposure to pesticides used on trees near the parking lot to control aphids.[157]

Specifically, the bill would have required EPA to "suspend the registration of imidacloprid, clothianidin, thiamethoxam, dinotafuran, and any other members of the nitro group of neonicotinoid insecticides to the extent such insecticide is registered, conditionally or otherwise," under FIFRA (7 U.S.C. 136 et seq.) "for use in seed treatment, soil application, or foliar treatment on bee attractive plants, trees, and cereals."

These suspensions on existing registrations and restrictions on new registrations would remain until more research is conducted and EPA determines that the insecticide will not cause "unreasonable adverse effects" on pollinators, including honey bees, native bees, and other pollinators. Such an assessment would be based on an evaluation of the published and peer-reviewed scientific evidence and a completed field study. The bill also would have required the Department of the Interior to coordinate with EPA in monitoring the health and populations of native bees, and annually report to Congress on their health and population status.

H.R. 2692 was similar to proposals being implemented or considered in the EU and Canada. In particular, the bill's language regarding "bee attractive plants, trees, and cereals" is similar to that under the EU's ban of the use of three neonicotinoid pesticides—clothianidin, imidacloprid and thiamethoxam—for "seed treatment, soil application (granules) and foliar treatment on bee attractive plants and cereals."[158] However, the EU's restrictions provide for certain exceptions (such as greenhouses and open-air fields after flowering), which are not provided for in H.R. 2692. (See "Restrictions in the European Union" for additional information on the EU's ban.)

In September 2014, the sponsors of H.R. 2692, along with 58 other Members of the U.S. House of Representatives, sent a letter to EPA urging the agency to "restrict and/or suspend the use of neonicotinoids on bee-attractive crops and ornamental application," including restricting the "times, methods of application, and location" for use of these pesticides, as well as urging EPA to review its policies related to EPA's pesticide registration, among other recommendations.[159] EPA received a similar letter from many environmental and sustainable agriculture organizations, including CFS, Friends of the Earth, Food and Water Watch, and other groups.[160]

Groups opposed to restrictions on pesticide use contend that comprehensive studies "challenge unsubstantiated claims against pesticides as a significant cause of colony decline" and note that "infrequent accidental exposures are not indicative of the general health of honey bee colonies."[161] Some further claim "there is no evidence linking neonicotinoids to bee declines" or to adverse health effects on bees.[162] They generally challenge studies that target neonicotinoids as the sole factor contributing to pollinator declines, and further maintain that unwanted pesticide exposure may be best addressed through "effective product labeling and the implementation of meaningful stewardship actions that help minimize harmful interactions."[163] Others question whether crop-applied pesticides pose a major risk to bees, given current approved uses and beekeeping practices.[164]

Expedited Pesticide Registration for Managing Parasitic Pests

In the 113th Congress, Representative Austin Scott introduced H.R. 5447, which would have amended U.S. pesticide laws (FIFRA; 7 U.S.C. 136a(c)(10)) to provide for expedited registration of pesticides that "improve managed pollinator bee health, including managing resistance to parasitic pests" and for expedited review of a pesticide registration that is "reasonably expected to improve the health of managed pollinator bees, including managing resistance to parasitic pests of managed pollinator bees," such as miticides. H.R. 5447 would have also required reports to Congress from both USDA and EPA. The report required from USDA would address the "extent and scope of the threat to the health of managed pollinator bees" from pathological factors (such as the parasitic mite, *Varroa destructor*; other arthropod pests; and fungal, microbial, and viral diseases) and from environmental factors (including habitat, forage, beekeeper practices and husbandry, and nutritional needs of managed bees). The report required by EPA would address the availability of pesticides to manage parasites and also EPA's efforts to expedite approvals of new products to control parasites of managed bees. The bill would have defined "managed pollinator bee" to mean "any bee that is raised and housed in a managed hive or other appropriate housing and used for honey production, managed pollination of crops, or breeding for commercial purposes." The bill was supported by some U.S. crop producers and other industry groups.

Some advocacy groups were opposed to H.R. 5447 and contend that "fast-tracking pesticide approvals" has contributed to current concerns involving bees and pesticides.[165] They further complain that focusing attention on other factors contributing to bee declines (such as mites) tends to shift attention away from remedies that could address beekeeper concerns about pesticide exposure, such as instituting agricultural best management practices (BMPs) that might avoid application of insecticides during bloom, and minimize exposure to bees by avoiding applications when pollen is available and bees are feeding.[166]

In February 2015, EPA announced its intentions to fast track approval of oxalic acid dihydrate for use as a miticide in beehives.[167] Oxalic acid has been used against varroa mites and is generally considered more safe compared to other higher risk chemicals. It has been approved for use in Canada since 2010, where a government analysis there claims: "An evaluation of available scientific information found that under the approved conditions of use the product has value and does not present an unacceptable risk to human health or the environment."[168]

APPENDIX. REVIEW OF THE SCIENTIFIC LITERATURE[169]

Following is a summary of selected scientific literature based largely on publications in peer-reviewed journals.

Bee Exposure to Pesticides

Researchers at USDA and others found that "high levels" of fungicides were present in both crop and non-crop pollen collected by bees, and that "real world pollen-pesticide blends affect honey bee health."[170] These scientists found very high levels of the fungicide chlorothalonil in pollen and wax,[171] and found high levels of at least four insecticides, three other fungicides, and an herbicide. Chlorothalonil and the two miticides, fluvalinate and coumaphos, were most frequently detected in pollen and wax. A 2010 study tested samples of beebread,[172] trapped pollen,[173] brood nest wax,[174] beeswax foundation,[175] and adult bees and brood and found a broad range of pesticides, including acaricides (which kill arachnids like spiders, ticks, and mites), fungicides, insecticides, and herbicides.[176]

Bees can be exposed to neonicotinoids in many ways. One study identified multiple routes of exposure to low levels of neonicotinoids for honey bees living and foraging near agricultural fields planted with corn or soybeans.[177] The highest potential exposure to the pesticides appeared to occur during planting season, when bee mortality was also high. Clothianidin was found in about half the bee-collected pollen sampled, thiamethoxam (which is quickly metabolized to become clothianidin) in 3 of 20 samples, and fungicides in all pollen samples. Levels of clothianidin in some pollen were high enough to kill bees. Clothianidin was detected in all dead and dying bees but in no healthy bees. Although corn is not an insect-pollinated crop, this research demonstrated that bees forage for corn pollen and take it back to the hive. About half the hive pollen sampled came from corn in this study. In addition, the study documented high levels of pesticide in exhaust material from mechanical planters when pesticides and talc were used to coat seeds, and found clothianidin in soil samples from planted and unplanted fields.

Other studies have measured neonicotinoids in pollen and nectar of canola (rape seed),[178] corn,[179] and sunflowers[180] grown from seed coated with pesticides. Levels found were below those known to be acutely toxic. Other studies have found that imidacloprid and thiamethoxam concentrations in nectar were greater in squash[181] and pumpkin flowers[182] when insecticide was applied to the soil than they were in canola and sunflowers grown from seed treated with neonicotinoids. Researchers also found metabolites of imidacloprid and thiamethoxam (clothianidin) in all parts of squash plants, along with the parent compound.[183]

A 2012 study found levels of neonicotinoids in bee-collected corn pollen that were similar to levels of imidacloprid determined by other scientists to have sublethal effects potentially affecting colony health.[184] Similarly, a subsequent study in 2013 found that imidacloprid levels in pollen gathered from bees in the field were high enough to cause sublethal effects on honey bees and bumble bees, based on laboratory research.[185] A 2013 study by the European Food Safety Agency (EFSA) of three neonicotinoid pesticides—imidacloprid, clothianidin, and thiamethoxam— determined that bees faced several risks including exposure to pollen and nectar, dust, and guttation fluid from maize.[186]

Another potential source of exposure was suggested by another 2012 study, showing that bees might be exposed to neonicotinoids in the corn syrup they are sometimes fed during the winter by beekeepers.[187] However, the researchers did not sample corn syrup actually fed to bees, but rather showed that bees would consume sufficient imidacloprid to produce toxic sublethal effects if they were provided contaminated corn syrup in the hive. Other research found no pesticides in samples of high-fructose corn syrup obtained from three major suppliers.[188]

Pesticide Effects on Bee Health

According to research cited in a 2007 study by the National Research Council (NRC), "the application of pesticides, especially insecticides used to control crop pests, kills or weakens thousands of honey bee colonies in the United States each year."[189] Nevertheless, the study concluded that *local* bee kills "likely have not contributed significantly to the recent *national* decline in colony populations" [emphasis added].[190]

Recent studies have begun to identify mechanisms by which some chemical interactions occur.[191] For example, some fungicides may inhibit an enzyme that bees need to detoxify miticides.[192] However, further research is needed since measurement difficulties continue to complicate study of this issue, including difficulty testing the in-field or field-realistic doses; effects at different life stages; how to account for cases of accidental exposure or exposure to multiple different types of pesticides; and how to account for species diversity (e.g., between honey bees and native bees), differences in nutritional needs/access and nesting sites, biological organization, and floral specialization.[193]

Generalizations about the relative importance of pesticides for global bee health cannot be drawn from available data, given the disparate study designs and results. Moreover, research has been and continues to be conducted, with most scientists focusing on a single pesticide or pathogen at a time. Consideration of interactions has been minimal—whether exposure from different types of chemicals or exposure from chemicals in combination with other factors. One study explains:

> Attempts to correlate global bee declines or CCD with increased pesticide exposures alone, have not been successful to date.... Pesticide interactions among various mixtures as well as with other stressors including *Varroa* and *Nosema*, IAPV, beneficial hive microbes, and impacts on bee immune systems all require further study. It seems to us that it is far too early to attempt to link or to dismiss pesticide impacts with CCD.[194]

The study noted that the doses of individual pesticides found in bees were not concentrated enough to be lethal, but the authors remained concerned about possible chronic problems caused by long-term exposure and possible additive or synergistic effects of exposure to the combinations of pesticides found.[195] For example, numerous pesticides classified as pyrethroids that were found in the field have been shown to cause disorientation in honey bees.[196]

It is important to note that virtually all information about pesticide risks to bees derives from studies in cultivated fields or laboratories. There is little or no information about the possible impacts on bee populations of pesticides applied by homeowners (for example, to control mosquitos) or to ornamental plantings.

Nevertheless, a sizeable and growing body of scientific research presents compelling evidence suggesting that pesticide exposure may be harmful to bees and other animal species. For example, a 2014 meta-analysis covering 800 peer-reviewed reports over two decades examined the reported effects of systemic pesticides on ecosystems and a range of animal species—terrestrial invertebrates, including insect pollinators; aquatic invertebrates; and birds.[197] The analysis, conducted by entomologists and ornithologists based in Europe, concluded that neonicotinoid pesticides "pose a serious risk to honeybees and other pollinators such as butterflies and to a wide range of other invertebrates such as earthworms and vertebrates including birds."[198]

In response to reports that honey bees are disappearing and causing hives to collapse, recent studies of the impacts of exposure to imidacloprid and other neonicotinoids[199] have focused more on their potential to affect complex behaviors in insects, including flight, navigation, olfactory memory, recruitment, foraging, and coordination.[200] One study reported sublethal effects of neonicotinoid pesticides on honey bee foraging behavior that may impair the navigational and foraging abilities of honey bees.[201] Another study found a reduction of foraging activity and homing behavior of honey bees exposed to treated crops at certain exposure levels of neonicotinoids.[202]

Scientists at Bayer Crop Science argue with the findings of these and other studies, claiming that the dose of pesticide delivered to bees in some studies is not "field-relevant," resulting in findings obtained "under artificial conditions" and "in conflict with" earlier studies.[203] Some of the statements made in the 2012 Bayer Crop Science review of the scientific literature have been rebutted by researchers at Purdue University and Pennsylvania State University, based on comments submitted by the Center for Food Safety as part of EPA's public rulemaking docket.[204] Among a list of concerns regarding various statements in the Bayer Crop Science report, Professor Frazier from Pennsylvania State University states: "To object to a specific dose on the basis of an estimated level without giving the reader the actual basis for this neglects not only good science protocol, but also obfuscates any arguments."[205] Among other comments, Professor Krupke from Purdue University further acknowledges: "These data suggest that there is a strong likelihood that the neonicotinoids thiamethoxam and/or clothianidin were responsible for the bee kills we investigated."[206]

EPA has determined that clothianidin "has the potential to be highly toxic on both a contact and an oral basis" to honey bees.[207] EPA also has reported that one honey bee field study submitted to the agency indicates that "mortality, pollen foraging activity, and honey yield were negatively affected by residues of clothianidin," but the residue levels causing the effect were not reported. Acute effects also have been demonstrated in another field study. It showed that honey bees can be killed by exposure to pesticide-contaminated talc if they fly through dust clouds associated with planting,[208] but mortality appears to depend on high levels of humidity.[209] Health Canada has concluded that corn planting also was implicated in bee mortalities in Ontario during the spring planting season in 2012.[210]

Research has shown that honey bee exposure to sublethal levels of pesticides, including neonicotinoids, exhibit impaired brood development,[211] impaired olfactory associative behavior,[212] and impaired homing ability.[213] Exposure to sublethal levels of neonicotinoid has also been associated with higher rates of *Nosema* infection,[214] and also reduced immune functioning in those bees infected,[215] making bees more susceptible to viral infections.[216] In another study, responsiveness to sucrose and the bee's "waggle dancing" abilities[217] were adversely affected when honey bees ingested imidacloprid.[218] Another study alternatively fed honey bee hives corn syrup treated with imidacloprid, while other hives were fed untreated corn syrup. After 6 months, nearly all the treated hives collapsed, while the untreated control hives remained healthy.[219] Another study showed that imidacloprid ingestion by larvae of the stingless bee (*Melipona quadrifasciata anthidioidides*) resulted in decreased survival rates, negatively affected development of a specific region of the bee brain, and impaired walking behavior of newly emerged adult worker bees.[220]

A 2013 study found that exposure to a combination of an insecticide (imidacloprid) and a miticide (coumaphos) impaired learning and memory formation, important behaviors involved in foraging, in honey bees exposed under "field-realistic concentrations."[221] A similar study also showed cognitive impairments from exposure to these same pesticides.[222] Studies also have found increased honey bee susceptibility to and mortality from *Nosema*[223] after exposure to sublethal concentrations of some pesticides (fipronil, amitraz, fluvalinate, chlorothalonil, pyraclostrobin, and imidacloprid).[224] Another study concluded that high levels of the fungicide chlorothalonil in pollen and wax may be associated with entombing behavior,[225] a sign that a hive has been poisoned. These and other pesticides commonly used to control mites have also been shown to have toxic effects on developing honey bee larvae at levels currently found in hives.[226]

A 2014 study reported linkages between CCD and sublethal exposure of neonicotinoid pesticides—imidacloprid and clothianidin—which affected the winterization of the bees, comparing exposed bee colonies to a control of non-exposed bee colonies.[227] In the study, one-half the bee colonies exposed to neonicotinoids exhibited symptoms resembling CCD and had abandoned their hives during the winter, whereas one-sixth of the control colonies were lost exhibiting *Nosema ceranae* symptoms. The control colonies did not abandon their hives but repopulated quickly; the surviving exposed colonies were small and either without queen bees or had no brood, according to the study. Some have criticized this study for its small sample size, lack of replication at each location, and the use of pesticide doses that may not reflect realistic field conditions.[228]

Researchers at USDA and affiliated bee laboratories at land grant universities throughout the United States continue to study the effects of neonicotinoids and other pesticides, including the effects of interactions between pesticides and other stressors.[229] Some of these studies demonstrate that exposure by bees to certain pesticides and/or combinations of insecticides (e.g., neonicotinoids with miticides) may affect the bee's overall energy budget, including its metabolism, physical activity, digestion, and immunity; cause other types of physiological effects; and affect different parts of the bee's nervous system, depending on the type or combination of insecticides studied among other variables, including diet and nutrition.[230]

A few studies have examined the effects of pesticides on various species of bumble bees. One study found that bumble bee exposure to imidacloprid may affect bee brain functioning, as well as colony growth and nest conditions.[231] Another study found that bumble bee colonies exposed in a laboratory to low levels of imidacloprid had a significantly reduced growth rate and an 85% reduction in queen production relative to untreated colonies.[232] Other studies have found that exposure of neonicotinoids could adversely affect foraging behavior,[233] reproductive success,[234] and locomotor behavior.[235] Another study showed effects following direct contact with a range of pesticides, including shortened life spans and inability to produce brood in worker bees; however, some pesticides and all fungicides tested reportedly did not result in sublethal effects.[236] Yet another study reported no adverse impacts on brood production or other sublethal effects on bumble bees from exposure to the neonicotinoid thiamethoxam.[237] One study showed that chronic exposure to two pesticides, one a neonicotinoid, at concentrations close to those found in fields "impairs natural foraging behavior and increases worker mortality leading to significant reductions in bumblebee brood development and colony success."[238] Bayer Crop Science argues that the results of some of these studies were also obtained "under artificial conditions and are in conflict with" earlier studies.[239]

A 2013 study examining the health of bumble bee colonies placed near crops treated with neonicotinoids were inconclusive due to weather and insufficient data, underlining the need to conduct further field studies that complement laboratory studies.[240] Some cite this study as evidence against the need to impose restrictions on the use of neonicotinoid pesticides,[241] while others criticize some of the study's published sources and research methodologies.[242]

End Notes

[1] More information on the difference between managed honey bees and native bees is available in CRS Report R43191, *Bee Health: Background and Issues for Congress.*

[2] United Nations Environment Programme (UNEP), *Global Honey Bee Colony Disorders and Other Threats to Insect Pollinators*, UNEP Emerging Issues, 2010.

[3] For more information, see CRS Report R43191, *Bee Health: Background and Issues for Congress.*

[4] Bee colony losses are common during the winter months, especially in colder climates. Hives should be strong and healthy before going into winter in order to minimize losses.

[5] D. vanEngelsdorp, et al., "Colony Loss 2013-2014," May 15, 2014, http://beeinformed.org/.

[6] K. Kaplan, "Fact Sheet: Survey of Bee Losses During Winter of 2012/2013," http://www.ars.usda.gov/is/br/beelosses/index.htm. Whether losses may be attributable to CCD may be based, in part, on reported colony losses "with no dead bees present, which is indicative of CCD." See USDA, *CCD Progress Report*, June 2012, p. 9.

[7] USDA, *Report on the National Stakeholders Conference on Honey Bee Health*, National Honey Bee Health Stakeholder Conference Steering Committee, May 2013, http://www.usda.gov/documents/ReportHoneyBeeHealth.pdf. This document, widely referred to as the USDA-EPA joint report, provides proceedings of an October 2012 workshop convened by USDA and EPA (edited by the National Honey Bee Health Stakeholder Conference Steering Committee).

[8] See, for example, USDA's annual CCD Research Progress Reports, available at http://www.ars.usda.gov/News/ docs.htm?docid=15572.

[9] USDA-EPA joint report, p. 16. Attributed to conferees Reed Johnson (Ohio State University) and James Frazier (Pennsylvania State University). The report references consensus by the Pesticide Risk Assessment for Pollinators in a 2011 Executive Summary, published by the Society of Environmental Toxicology and Chemistry (SETAC).

[10] FIFRA § 2(u). 7 U.S.C. 136-136y. See also CRS Report RL31921, *Pesticide Law: A Summary of the Statutes.*

[11] S. Kaiser, EPA, personal communication with CRS staff, December 16, 2011.

[12] 7 U.S.C. §136 et seq.

[13] 21 U.S.C. §§ 301, et seq.

[14] The Federal Seed Act (7 U.S.C. § 1561) defines a treated seed as any seed "given an application of a substance or subjected to a process designated to reduce, control, or repel disease organisms or other pests, which attack seeds or seedlings growing therefrom."

[15] For more information, see CRS Report RL31921, *Pesticide Law: A Summary of the Statutes*; and CRS Report 96- 759, *Pesticide Legislation: Food Quality Protection Act of 1996 (P.L. 104-170).*

[16] EPA, "About Pesticide Registration," http://www2.epa.gov/pesticide-registration/about-pesticide-registration. Other information is available at CRS Report RL32218, *Pesticide Registration and Tolerance Fees: An Overview.*

[17] UNEP, *Global Honey Bee Colony Disorders and Other Threats to Insect Pollinators*, 2010.

[18] S. P. Bradbury, Office of Pesticide Programs, letter to Peter T. Jenkins, Center for Food Safety and International Center for Technology Assessment, July 17, 2012. Hereafter referred to as Bradbury response to citizen petition.

[19] See, for example, D. Goulson, "An Overview of the Environmental Risks Posed by Neonicotinoid Insecticides," *Journal of Applied Ecology*, vol. 50, issue 4 (2013), pp. 977–987; E. Stokstad, "Pesticides Under Fire for Risks to Pollinators," *Science*, v. 340, no. 6133 (May 10, 2013), pp. 674-676; and D. Gibbons, C. Morrisey, P. Mineau, "A Review of the Direct and indirect Effects of Neonicotinoids and Fipronil on Vertebrate Wildlife," *Environ. Sci. Pollut. Res.*, June 2014.

[20] D. vanEngelsdorp and M. D. Meixner, "A Historical Review of Managed Honey Bee Populations in Europe and the United States and Factors That May Affect Them," *Journal of Invertebrate Pathology*, v. 103 (2010), Supp. 1, pp. S80- S95.

[21] L. Maxim and J. P. van der Sluis, "Expert Explanations of Honeybee Losses in Areas of Extensive Agriculture in France: Gaucho® Compared with Other Supposed Causal Factors," *Environmental Research Letters*, v. 5 (2010).

[22] D. Laurino, M. Porporato, A. Patetta, et al., "Toxicity of Neonicotinoid Insecticides to Honey Bees: Laboratory Tests," *Bulletin of Insectology*, v. 64, n. 1 (2011), pp. 107-113.

[23] Pyrethroids are synthetic pesticides modeled on the naturally occurring pyrethrins that are formed by chrysanthemums. Most pyrethroids are more toxic and persistent than pyrethrins. Fenvalerate (an insecticide), bifenthrin (a miticide and an insecticide), cypermethrin (an insecticide), and fluvalinate (a miticide) are pyrethroids.

[24] J. L. Frazier, presentation to Senate staff on pesticides and bee health, April 17, 2013.

[25] Pesticides broadly refer to chemicals used to prevent, destroy, or repel pests, such as insecticides, herbicides, fungicides, rodenticides, and miticides, among others. Pesticides also are used to kill organisms that can cause diseases.

[26] Some common pesticides applied deliberately to hives include formic acid (an organic acid), thymol (an essential oil), coumaphos (an organophosphate), and fluvalinate (a pyrethroid).

[27] A 2012 study looked at the impact on bee health of three categories of inerts: nonionic surfactants, crop oil concentrates, and organosilicone surfactants. Bee behavior was adversely affected after ingesting organosilicone surfactant. Nonionic adjuvants also had a small effect, while the crop oil concentrates were inactive See T. J. Ciarlo, C. A. Mullin, James L. Frazier, et al., "Learning Impairment in Honey Bees Caused by Agricultural Spray Adjuvants," *PLoS ONE*, v. 7, n. 7 (2012), p. e40848.

[28] National Pesticide Information Center, Piperonyl Butoxide (fact sheet), http://npic.orst.edu/factsheets/pbogen.pdf.

[29] See, for example, C. H. Krupke, G. J. Hunt, B. D. Eitzer, et al., "Multiple Routes of Pesticide Exposure for Honey Bees Living Near Agricultural Fields," *PLoS ONE*, v. 7. n. 1 (January 2012), p. e29268.

[30] USDA-EPA joint report, p. 17.

[31] J. H. Cane and V. J. Tepedino, "Causes and Extent of Declines Among Native North American Invertebrate Pollinators: Detection, Evidence, and Consequences," *Conservation Ecology*, v. 5, n. 1 (2001), p. 1.

[32] Pesticides other than insecticides, such as fungicides or miticides, may be toxic to bees, although bees are insects and not the targeted species.

[33] *Neonicotinoid Pesticides and Honey Bees*, Washington State University Extension, Fact Sheet FS122E, http://cru.cahe.wsu.edu/CEPublications/FS122E/FS122E.pdf; and comments by Nathalie Steinhauer (University of Maryland) to the Virginia Native Plant Society, October 9, 2014.

[34] For example, see J. Hopwood, et al., *Are Neonicotinoids Killing Bees? A Review of Research into the Effects of Neonicotinoid Insecticides on Bees, with Recommendations for Action*, 2012, and *A Review of Research into the Beyond the Birds and the Bees: Effects of Neonicotinoid Insecticides on Agriculturally Important Beneficial Invertebrates*, 2013 (both prepared for Xerces Society for Invertebrate Conservation); V. Kindemba, *The Impact of Neonicotinoid Insecticides on Bumblebees, Honey Bees, and Other Non-Target Invertebrates,* 2009 (Invertebrate Conservation Trust); R. Heintzelman, et al., "Overview of Recent Publications on Neonicotinoids and Pollinators," May 2012 (prepared for Bayer Crop Science); Center for Food Safety (CFS), "Pollinators and Pesticides: A Report by Center For Food Safety on Pollinator Health, Research, and Future Efforts for Pollinator Protection," September 2013; CFS, "Heavy Costs: Weighing the Value of Neonicotinoid Insecticides in Agriculture," March 2014; CFS, Pollinator Study Index (available at http://www.centerforfoodsafety.org/files/pollinator-study-index_73710.pdf); M. Simon, *Follow the Honey: 7 Ways Pesticide Companies are Spinning the Bee Crisis to Protect Profits*, April 2014; and PANNA, *Pesticides and Honey Bees: State of the Science*, May 2012.

[35] For more information, see P. Mineau and C. Palmer, *The Impact of the Nation's Most Widely Used Insecticides in Birds*, American Bird Conservancy, March 2013; and C. A. Hallmann, et al., "Declines in Insectivorous Birds Are Associated with High Neonicotinoid Concentrations," *Nature*, vol. 511 (July 17, 2014), pp. 341-343.

[36] Educational resources are available from most land grant universities and also the following organizations: Bee Informed Partnership (www.beeinformed.org); Center for Urban Ecology and Sustainability/Pollinator Conservation (www.entomology.umn.edu/cues /pollinators); National Pesticide Information Center (www.npic.orst.edu); Pesticide Environmental Stewardship (www.pesticidestewardship.org); and Pollinator Partnership (www.pollinator.org).

[37] PES, "Pollinator Protection," http://pesticidestewardship.org/PollinatorProtection/Pages /default.aspx.

[38] EPA, "Find Best Management Practices to Protect Pollinators," http://www2.epa.gov /pollinator-protection/find-bestmanagement-practices-protect-pollinators.

[39] C. H. Krupke, G. Hunt, and R. E. Foster, "Protecting Honey Bees from Pesticides," Purdue University Extension, E-53-W, http://extension.entm.purdue.edu/publications/E-53.pdf.

[40] Ibid.

[41] Almond Board of California, "Honey Bee BMPs," http://www.almonds.com/newsletters /outlook/honey-bee-bmpspest-management-and-honey-bees. See also testimony of Arthur

Cummings, Capay Farms, at a House Agriculture Subcommittee on Horticulture, Research, Biotechnology, and Foreign Agriculture hearing, April 29, 2014.

[42] W. Buhler, "Pesticide Applicator BMPs," Pesticide Environmental Stewardship, available at http://pesticidestewardship.org/PollinatorProtection/PesticideApplicatorBMPs/Pages/Pesticide-Applicator-BMPs.aspx.

[43] Bayer pamphlet, "Honey Bee Health: Understanding the Issues, Providing Solutions."

[44] See, for example, E. Lichtenberg, D. Parker, and S. Lane, "Best Management Practice Use and Nutrient Management in Maryland: A 2010 Snapshot," 2010, University of Maryland; D. Stoddard, "Monitoring, Prevention and Mitigation of Agricultural Contaminant Sources," November 2006; Minnesota Department of Agriculture; M. Ribaudo, J. Savage, M. Aillery, *An Economic Assessment of Policy Options To Reduce Agricultural Pollutants in the Chesapeake Bay*, June 2014; and American Farmland Trust, "The Adoption of Conservation Practices in Agriculture," August 2013.

[45] CRS communication with the Center for Food Safety, February 13, 2015.

[46] Health Canada, "Notice of Intent, NOI2013-01, Action to Protect Bees from Exposure to Neonicotinoid Pesticides."

[47] Honey Bee Health Coalition, "Bee Healthy Roadmap," October 2014, https://www.keystone.org/images/keystonecenter/spp-documents/Environment/BeeHealth/Bee-Healthy-Roadmap-October-2014.pdf. Also testimony of David Fischer, Bayer North American Bee Care Center, at a House Agriculture Subcommittee on Horticulture, Research, Biotechnology, and Foreign Agriculture hearing, April 29, 2014.

[48] CRS communication with the Center for Food Safety, February 13, 2015 (citing D. Goulson, "An Overview of the Environmental Risks Posed by Neonicotinoid Insecticides," *Journal of Applied Ecology*, vol. 50, issue 4 (2013), pp. 977–987.

[49] See, for example, Washington State Department of Agriculture (WSDA), "10 Ways to Protect Bees from Pesticides," AGR PUB 701-388, http://agr.wa.gov/fp/pubs/docs/388-TenWaysToProtectBeesFromPesticides.pdf.

[50] Ibid.

[51] See, for example, Xerxes, "Protecting Bees from Neonicotinoid Insecticides in Your Garden," 2013; J. Hopwood and M. Shepherd, "Neonicotinoids in Your Garden," 2012; and Friends of the Earth, *Gardeners Beware: Bee-Toxic Pesticides Found in "Bee-Friendly" Plants*, June 2014 (updated).

[52] For a listing of these products, see Center for Food Safety, "Help Save the Bees," April 2013, http://www.centerforfoodsafety.org/files/pesticide_list_final_59620.pdf.

[53] White House, "Presidential Memorandum—Creating a Federal Strategy to Promote the Health of Honey Bees and Other Pollinators," June 20, 2014, http://www.whitehouse.gov/the-press-office/2014/06/20/presidential-memorandumcreating-federal-strategy-promote-health-honey-b.

[54] Announcement by Michael Stebbins, White House Office of Science and Technology Policy, at the 14th annual conference of the North American Pollinator Protection Campaign (NAPPC), October 22, 2014. Other information is available in CRS Report R43191, *Bee Health: Background and Issues for Congress.*

[55] See, for example, letter to U.S. President Barack Obama from more than 100 environmentally friendly food companies urging the Administration to protect bees and other pollinators from pesticides, January 13, 2015.

[56] J.R. Pegg, "Neonicotinoid Controversy Hits Pollinator Health," *Food Chemical News*, December 3, 2014.

[57] EPA, Pollinator Protection Strategic Plan, http://www2.epa.gov/pollinator-protection/pollinator-protection-strategicplan.

[58] USDA, *Report on the National Stakeholders Conference on Honey Bee Health*, National Honey Bee Health Stakeholder Conference Steering Committee, May 2013, http://www.usda.gov/documents/ReportHoneyBeeHealth.pdf.

[59] EPA news release, "USDA and EPA Release New Report on Honey Bee Health," May 2, 2013.

[60] More information is available in CRS Report R43191, *Bee Health: Background and Issues for Congress.*

[61] EPA, "Groups of Pesticides in Registration Review," http://www2.epa.gov/pesticide-reevaluation/groups-pesticidesregistration-review.

[62] NRDC, "Petition for Interim Administrative Review of Neonicotinoid Pesticides," filed July 7, 2014, http://docs.nrdc.org/health/files/hea_14070701a.pdf.

[63] Public statements by EPA's Assistant Administrator for the Office of Chemical Safety and Pollution Prevention, Jim Jones, at the 14th Annual NAPPC International Conference, October 22, 2014.

[64] FIFRA section 3(c)(5).

[65] FIFRA section 3(c)(7).

[66] For more information, see EPA's website: http://www2.epa.gov/pesticide-registration /conditional-pesticideregistration.

[67] A pesticide's labels and the indication of their current conditional (or not) status at EPA's Label database:EPA, "Pesticide Product Label System," http://iaspub.epa.gov/apex /pesticides/f?p=PPLS:1.

[68] See, for example, GAO, *EPA Should Take Steps to Improve Its Oversight of Conditional Registrations*, GAO-13- 145, August 2013; and GAO, *EPA's Formidable Task To Assess and Regulate Their Risks*, RCED-86-125: April 1986.

[69] J. Sass and M. Wu, "Superficial Safeguards: Most Pesticides Are Approved by Flawed EPA Process," *NRDC Issue Brief*, IB: 13-01-B, March 2013, http://www.nrdc.org/health /pesticides/files/flawed-epa-approval-process-IB.pdf.

[70] Ibid.

[71] See, for example, GAO, *EPA Should Take Steps to Improve Its Oversight of Conditional Registrations*, GAO-13- 145, August 2013.

[72] EPA, Environmental Fate and Effects Division (EFED), EFED Response to Comments Submitted to the Clothianidin Registration Review Docket (docket: EPA-HQ-OPP-2011-0865), June 11, 2012. References EPA's interim guidance from 2011, *Pesticides: Science and Policy, Interim Guidance on Honey Bee Data Requirements.*

[73] Meeting materials and public comments are posted in the regulatory docket (docket: EPA-HQ-OPP-2012-0543). EPA, "Proposed Decision Framework for Assessing Risks of Pesticides to Honey Bees," presentation at the 2012 FIFRA SAP meeting. Also see EPA's presentation, "Ecological Risk Assessment Process for Honey Bees."

[74] EPA, *Guidance for Assessing Pesticide Risks to Bees*, June 19, 2014, http://www2.epa.gov/sites/production/files/2014-06/documents/pollinator_risk_assessment_guidance_06_19_14.pdf.

[75] CRS communication with the Center for Food Safety, February 13, 2015.

[76] USDA, "Enhancing Nest Sites For Native Bee Crop Pollinators," *Agroforestry Notes*, AF Note 34, September 2007.

[77] EPA, "Benefits of Neonicotinoid Seed Treatments to Soybean Production," October 15, 2014, http://www2.epa.gov/ sites/production/files/2014-10/documents/benefits_of_neonicotinoid_ seed_treatments_to_soybean_production_2.pdf.

[78] EPA, "Benefits of Neonicotinoid Seed Treatments to Soybean Production," October 15, 2014. See also http://www2.epa.gov/pollinator-protection/benefits-neonicotinoid-seed-treatments-soybean-production

[79] Public statements by EPA's Assistant Administrator for the Office of Chemical Safety and Pollution Prevention, Jim Jones, at the 14th Annual NAPPC International Conference, October 22, 2014.

[80] EPA, "Summit on Reducing Exposure to Dust from Treated Seed," http://www2.epa.gov/pollinator-protection/2013- summit-reducing-exposure-dust-treated-seed.

[81] 79 *Federal Register* 172: 53009: September 5, 2014.

[82] Stover refers to the leaves and stalks of field crops, such as corn or soybeans, commonly left in a field after harvest.

[83] T. Stecker, "Syngenta Asks EPA to Raise Tolerance Level for 'Bee-Killing' Chemical," *Greenwire*, September 5, 2014.

[84] EPA, "New Pesticide Labels Will Better Protect Bees and Other Pollinators," press release, August 15, 2013; and letter from EPA's Office of Investigations to the Pollinator Stewardship Council, February 28, 2014. See also letters from EPA to registrants of nitroguanidine neonicotinoid products, July 22, 2013, and August 15, 2013.

[85] EPA, "New Labeling for Neonicotinoid Pesticides," http://www2.epa.gov/pollinator-protection/new-labelingneonicotinoid-pesticides.

[86] J. R. Pegg, "State Pesticide Officials, EPA Ready to Join Forces on Pollinator Protection," *Food Chemical News*, December 10, 2014.

[87] For more information, see Pollinator Stewardship Council, "Analysis of the New Label," http://pollinatorstewardship.org/wp-content/uploads/2014/02/PSC-label-layout-of-concerns.pdf.

[88] Letter from EPA to registrants of nitroguanidine neonicotinoid products, August 15, 2013.

[89] Pollinator Stewardship Council, http://pollinatorstewardship.org/wp-content/uploads/2014/02 /PSC-label-layout-ofconcerns.pdf. See also blog posting, March 21, 2014, http://pollinatorstewardship.org/?p=2143.

[90] SFIREG Full Committee meeting, December 8, 2014, http://aapco.org/meetings/agen-120814.pdf.

[91] C. Purdy, "EPA to Review State Pollinator Protection Plans," *Politico Morning Agriculture*, December 9, 2014.

[92] J.R. Pegg, "State Pesticide Officials, EPA Ready to Join Forces on Pollinator Protection," *Food Chemical News*, December 10, 2014.

[93] EPA, "Introduction to Pesticide Drift," http://www2.epa.gov/reducing-pesticide-drift/introduction-pesticide-drift. Pesticide droplets are produced by spray nozzles used in application equipment for spraying pesticides on crops, forests, turf, and home gardens. Dust formulations refer to very fine dry particles of pesticides.

[94] EPA, "Reducing Pesticide Drift," http://www2.epa.gov/reducing-pesticide-drift.

[95] Information is at DriftWatch's website: https://www.driftwatch.org/.

[96] FWS memorandum, "Use of Agricultural Practices in Wildlife Management in the National Wildlife Refuge Service," July 17, 2014, http://www.peer.org/assets/docs/fws/FWS_ Memorandum.pdf.

[97] T. Stecker, "FWS Bans GMO Feed, Bee-Killing Chemicals Nationwide," *Greenwire*, August 1, 2014.

[98] See Region 1 FWS memorandum, "Guidelines Regarding the Interim Use and Phase Out of Neonicotinoid Insecticides to Grow Agricultural Crops for Wildlife on NWRs in the Pacific Region," June 9, 2014.

[99] OpEd by U.S. Representative Austin Scott, "Protecting Bee Health Is a Long-Term Commitment," *Agri-Pulse*, 2013.

[100] See, for example, Minnesota Department of Agriculture Pesticide and Fertilizer Management Division, *Scoping a Review of Neonicotinoid Use, Registration and Insect Pollinator Impacts in Minnesota*, March 2014 (draft).

[101] J.R. Pegg, "State Pesticide Officials, EPA Ready to Join Forces on Pollinator Protection," *Food Chemical News*, December 10, 2014.

[102] COLOSS (Prevention of honeybee COlony LOSSes), http://www.coloss.org/, collects data from 19 European countries, and also from Israel and Algeria.

[103] COLOSS, "Losses of Honey Bee Colonies over the 2013/14 Winter," July 18, 2014.

[104] COLOSS, "Losses of Honey Bee Colonies over the 2012/13 Winter," July 24, 2013.

[105] EpiloBee, "A Pan-European Epidemiological Study on Honeybee Colony Losses 2012-2013," August 2014. See also OPERA Research Center, *Bee Health in Europe—Facts & Figures* and *Bee Health in Europe, An Overview, 2012*, http://www.operaresearch.eu/.

[106] Regulation (EC) No 485/2013 of 24 May 2013 amending Implementing Regulation (EU) No 540/2011, as regards the conditions of approval of the active substances clothianidin, thiamethoxam and imidacloprid, and prohibiting the use and sale of seeds treated with plant protection products containing those active substances. See also European Commission, "Bees & Pesticides: Commission Goes Ahead with Plan to Better Protect Bees," http://ec.europa.eu/food/ archive/animal/liveanimals/bees/neonicotinoids_en.print.htm.

[107] Restrictions were placed in some member states, such as Italy (ban on corn seed coating), Slovenia (ban on corn and sugar beet seed coating), France (ban on corn and canola seed coating), and Germany (ban on corn seed coating). M. Dermine, "Neonicotinoids in the European Union," presentation at November 4, 2014, webinar.

[108] EFSA is an agency of the European Union that provides independent scientific advice and communication on existing and emerging risks associated with food and feed safety.

[109] EC, "Bee Health: What Is the EU Doing? Questions and Answers," April 7, 2014 http://europa.eu/rapid/pressrelease_MEMO-14-260_en.htm. See also EC press releases, "Bee Health: EU-Wide Restrictions on Pesticide Use to Enter into Force on 1 December," May 24, 2013; and "Bees and Pesticides: Commission to Proceed with Plan to Better Protect Bees," April 29, 2013.

[110] Regarding the EFSA study, EPA claims its own scientific conclusions "are similar to those expressed in the EFSA report with regard to the potential for acute effects and uncertainty about chronic risk," but notes "the EFSA report does not address risk management" which is a requirement for EPA under U.S. law. See EPA, "Colony Collapse Disorder: European Bans on Neonicotinoid Pesticides," August 2013, http://www.epa.gov/pesticides/about /intheworks/ccdeuropean-ban.html.

[111] EC, "Bee Health: EU Takes Additional Measures on Pesticides to Better Protect Europe's Bees," July 16, 2013, http://europa.eu/rapid/press-release_IP-13-708_en.htm. EFSA's analysis, "Conclusion on the Peer Review of the Pesticide Risk Assessment for Bees for the Active Substance Fipronil," *EFSA Journal 2013*, 11(5):3158, 27 May 2013, http://www.efsa.europa.eu/en/efsajournal/pub/3158.htm.

[112] See, for example, DEFRA, Food and Environment Research Agency (FERA), "Effects of Neonicotinoid Seed treatments on Bumble Bee Colonies Under Field Conditions," March 2013.

[113] See, for example, a 2013 study by the European Landowners' Organization (ELO), the European Crop Protection Association (ECPA), and the European Initiative for Sustainable Development in Agriculture (EISA), "Pollinators and Agriculture: Agricultural Productivity and Pollinator Protection."

[114] P. Case, "Neonicotinoid Field Studies to Assess Effect on Bees," *Farmers Weekly*, September 26, 2014. Reportedly, large-scale field experiments will look at the effects on bees of two seed treatments—clothianidin and thiamethoxam.

[115] See, for example, N. Robinson, "Neonic Ban Threatens UK Mustard Crop," *FoodManufacture*, August 19, 2014; and M McGrath, "Scientific opinions differ on bee pesticide ban," *BBC News*, April 23, 2013.

[116] CAPA, "CAPA Statement on Honey Bee Wintering Losses in Canada (2014)," July 2014. Losses in prior years are available at CAPA's website, http://www.capabees.com/.

[117] See, for example, Health Canada, "2012 Report on Pesticide Incidents," available at http://www.hc-sc.gc.ca/.

[118] Health Canada, "Evaluation of Canadian Bee Mortalities that Coincided with Corn Planting in Spring 2012," http://www.hc-sc.gc.ca/cps-spc/pubs/pest/_decisions/bee_corn-mort-abeille_mais/index-eng.php.

[119] Health Canada, "Ontario Bee Incidents 2012-Update," September 2013, http://www.hc-sc.gc.ca/cps-spc/pubs/pest/ _fact-fiche/bees-incidents-abeilles-2012/index-eng.php.

[120] Health Canada, "Notice of Intent, NOI2013-01, Action to Protect Bees from Exposure to Neonicotinoid Pesticides," September 2013, http://www.hc-sc.gc.ca/cps-spc/pest/part/consultations/_noi2013-01/noi2013-01-eng.php.

[121] Presentation by C. Scott-Dupree, "Field Study Reliability: A Honey Bee & Neonicotinoid Perspective," University of Guelph, http://sesss08.setac.eu/embed/sesss08/Cynthia_Scott-Dupree_Field_study_Reliability_-_A_Honey_bee__Neonicotinoid_PERSPECTIVE.pdf.

[122] Ibid.

[123] Health Canada, "Pollinator Protection: Reducing Risk from Treated Seed," April 2013, http://www.hc-sc.gc.ca/cpsspc/pubs/pest/_fact-fiche/pollinator-protection-pollinisateurs/index-eng.php.

[124] Health Canada, "Notice of Intent, NOI2013-01, Action to Protect Bees from Exposure to Neonicotinoid Pesticides."

[125] PMRA, "Re-evaluation Note REV2012-02, Re-evaluation of Neonicotinoid Insecticides," June 12, 2012, http://www.hc-sc.gc.ca/cps-spc/pubs/pest/_decisions/index-eng.php.

[126] See Ministry of the Environment and Climate Change, "Reducing Pesticide Use and Protecting Pollinator Health Ontario Sets Goal to Reduce Neonicotinoid Use by 80 Per Cent," November 25, 2014; and the Ontario Ministry of Agriculture, Food and Rural Affairs (OMAFRA), "Pollinator Health: A Proposal for Enhancing Pollinator Health and Reducing the Use of Neonicotinoid Pesticides in Ontario," November 2014.

[127] "Canadian Industry Seeks Broader Approach to Bee Problem," *Food Chemical News*, November 15, 2013.

[128] See, for example, Environmental Commissioner of Ontario, *Managing New Challenges*, Annual Report 2013/2014.

[129] Sun Parlor Honey Ltd. and Munro Honey v. Bayer CropScience Inc., Bayer Inc., Bayer AG, Syngenta Canada Inc. and Syngenta International AG, Court File No.: CV-14-21208 (Ontario Superior Court of Justice, Ontario, Canada), September 2, 2014. The complaint is

available at http://www.ontariobee.com/sites/ontariobee.com/files/DOCSLIB%232440628-v1-AMENDED_Claim.pdf.

[130] The lawsuit seeks C$400 million (about $370 million) in general and specific damages and C$50 million (about $46 million) in punitive damages, plus interest and legal costs.

[131] For example, CFS, "Heavy Costs: Weighing the Value of Neonicotinoid Insecticides in Agriculture" (March 2014) and "Pollinators & Pesticides" (September 2013); PANNA, "Honey Bees and Pesticides: State of the Science" (May 2012); and Friends of the Earth, "Follow the Honey" (April 2014) and "Gardeners Beware" (October 2013, June 2014).

[132] Steve Ellis, Tom Theobald, Jim Doan, Bill Rhodes, CFS, et al. v. Steven P. Bradbury and Bob Perciasepe, U.S. Environmental Protection Agency, Case No. 3:13-Cv-01266-Lb (U.S. District Court for the Northern District of California, San Francisco Division), May 31, 2013. In addition to CFS, the coalition is represented by Beyond Pesticides, Pesticide PANNA, Sierra Club, and the Center for Environmental Health, as well as four beekeepers with operations located in California, Colorado, Florida, New York, and Minnesota. The amended complaint is at http://www.centerforfoodsafety.org/issues/304/pollinators-and-pesticides/legal-actions. For more information, see CFS, "CFS, Beekeepers and Public Interest Groups Sue EPA over Bee-Toxic Pesticides," March 21, 2013, http://www.centerforfoodsafety.org/issues/304/pollinators-and-pesticides/press-releases.

[133] Pollinator Stewardship Council et al. v. USEPA and Dow Agrosciences, Case No.13-72346 (U.S. Court of Appeals for the 9th Circuit), December 13, 2013. The amicus curiae brief is at http://www.centerforfoodsafety.org/issues/304/ pollinators-and-pesticides/legal-actions.

[134] PANNA, CFS, and Beyond Pesticides v. California Department of Pesticide Regulation et al., Case No. RG14731906 (Superior Court for the State of California for the County of Alameda), July 8, 2014. The original complaint is at http://www.centerforfoodsafety.org /issues/304/pollinators-and-pesticides/legal-actions.

[135] Jeffrey Anderson, et al., v. State of Minnesota, Department of Natural Resources, A03-679 (State of Minnesota in Supreme Court), March 3, 2005 (posted: http://www.beyondpesticides.org/documents/MNSUPREME_Decision.pdf).

[136] Emergency Citizen Petition to the United States Environmental Protection Agency, March 12, 2013 http://www.centerforfoodsafety.org/files/cfs-clothianidin-petition-3-20-12_57213.pdf.

[137] EPA's response: S. P. Bradbury, EPA Office of Pesticide Programs, letter to Peter T. Jenkins, Center for Food Safety and International Center for Technology Assessment, "Clothianidin Emergency Citizen Petition dated March 20, 2012," July 17, 2012. The petition and numerous supplemental submissions of research reports and opinions are posted in the regulatory docket (docket: EPA-HQ-OPP-2012-0334-0015).

[138] 77 *Federal Register* 145: 44233-44234, July 27, 2012. Other information is in the regulatory docket (docket: EPAHQ-OPP-2012-0334).

[139] NRDC, "Petition for Interim Administrative Review of Neonicotinoid Pesticides," filed July 7, 2014, http://docs.nrdc.org/health/files/hea_14070701a.pdf.

[140] The June 20, 2013, letter is available at http://www.centerforfoodsafety.org/files/final-neonic-letter62013_43430.pdf. Also see P. Ambrosio, "Environmental Groups Urge President to Ban Neonicotinoids over Threat to Bees," *Bloomberg BNA Daily Environment Report*, June 24, 2013.

[141] See, for example, letter to U.S. President Barack Obama from more than 100 environmentally friendly food companies urging the Administration to protect bees and other pollinators from pesticides, January 13, 2015.

[142] "Crop Protection Industry Plans Response to EPA on Neonicotinoids," *Agri-Pulse*, October 29, 2014.

[143] Bayer pamphlet, "Honey Bee Health: Understanding the Issues, Providing Solutions."

[144] CropLife represents "companies that develop, manufacture, formulate and distribute crop protection chemicals and plant science solutions for agriculture and pest management" including Syngenta, Bayer, Dow, Dupont, and Monsanto.

[145] See, for example, Ag Infomatics, "Value of Neonicotinoids in North American Agriculture" (October 2014); and CropLife Foundation, "The Role of Seed Treatment in Modern U.S. Crop Production" (December 2013).

[146] For example: Bayer's "Bee Care Center" in North Carolina; Syngenta's funding of research grants to study causes of pollinator declines; and Monsanto's purchase in 2011 of the Beeologics research organization in Missouri.

[147] See, for example, postings by Corporate Europe Observatory, "Pesticides Against Pollinators: Private Letters Reveal Syngenta and Bayer's Furious Lobbying Against EU Measures to Save Bees," April 11, 2013, http://corporateeurope.org/agribusiness/2013/04/pesticides-against-pollinators.

[148] T. Rabesandratana, "Pesticidemakers Challenge E.U. Neonicotinoid Ban in Court," *Science*, August 28, 2013, http://news.sciencemag.org/europe/2013/08/pesticidemakers-challenge-e.u.-neonicotinoid-ban-court.

[149] Syngenta, "Syngenta Submits Legal Challenge to EU Suspension of Thiamethoxam," August 27, 2013, http://www.syngenta.com/global/corporate/en/news-center/news-releases/Pages /130827.aspx.

[150] T. Rabesandratana, "Pesticidemakers Challenge E.U. Neonicotinoid Ban in Court," *Science*, August 28, 2013, http://news.sciencemag.org/europe/2013/08/pesticidemakers-challenge-e.u.-neonicotinoid-ban-court.

[151] Syngenta, "Syngenta Submits Legal Challenge to EU Suspension of Thiamethoxam," August 27, 2013, http://www.syngenta.com/global/corporate/en/news-center/news-releases/Pages/130827.aspx.

[152] 79 *Federal Register* 172: 53009: September 5, 2014. For more information, see "EPA's Consideration to Modify Thiamethoxam Tolerance Levels."

[153] For more information, see CRS Report R43191, *Bee Health: Background and Issues for Congress.*

[154] See, for example, "Subcommittee Examines Research Efforts to Combat Pests and Diseases of Pollinators," House Agriculture Subcommittee on Horticulture, Research, Biotechnology, and Foreign Agriculture hearing, April 29, 2014.

[155] For example: "Impacts of Pesticides on Birds, Bees, and Broader Ecosystems" (March 2013); "Honey Bee Colony Health" (April 2013); "Applied Bee-Nomics" (June 2013); "What's Killing Our Bees?" (October 2013, December 2013); and "The Threat of Neonicotinoid Pesticides to Bees and Other Organism, and Risks to Human Health" (September 2014).

[156] T. Stecker, "New House Subcommittee Chief Will Push Forward on Bee Issues," *Environmental and Energy Policy*, December 16, 2014.

[157] See, for example, press release on the bill by Representative Earl Blumenauer's office.

[158] Regulation (EC) No 485/2013 of 24 May 2013 amending Implementing Regulation (EU) No 540/2011, as regards the conditions of approval of the active substances clothianidin, thiamethoxam and imidacloprid, and prohibiting the use and sale of seeds treated with plant protection products containing those active substances.

[159] Letter to EPA Administrator Gina McCarthy from Representatives Earl Blumenauer and John Conyers and 58 other House of Representatives Members, September 30, 2014.

[160] Letter to EPA Administrator Gina McCarthy from Representatives Earl Blumenauer and John Conyers and 58 other House of Representatives Members, September 24, 2014.

[161] Testimony of David Fischer, Bayer North American Bee Care Center, at a House Agriculture Subcommittee on Horticulture, Research, Biotechnology, and Foreign Agriculture hearing, April 29, 2014.

[162] Statements by CropLife representatives reported by J. R. Pegg, "Groups Press EPA, California to Protect Bees from Neonicotinoids," *Food Chemical News*, July 14, 2014; also written correspondence from David Fischer, Bayer CropScience LP, to members of the North American Pollinator Protection Campaign (NAPPC), November 18, 2014.

[163] Testimony of David Fischer, Bayer North American Bee Care Center, at a House Agriculture Subcommittee on Horticulture, Research, Biotechnology, and Foreign Agriculture hearing, April 29, 2014.

[164] See, for example, A Fairbrother et al., "Risks of Neonicotinoid Insecticides to Honeybees," *Environmental Toxicology and Chemistry*, vol. 33, no. 4 (April 2014).

[165] CFS press release, "New Pollinator Bill Helps Pesticide Industry, Not Bees or Beekeepers," September 12, 2014.

[166] Pollinator Stewardship Council, "House Subcommittee Forgot to Ask Beekeeping Industry," May 2, 2014.

[167] 80 *Federal Register* 23: 6084-6085, February 4, 2015.

[168] Canadian Honey Council, "Oxalic Acid," http://www.honeycouncil.ca/oxalic_acid.php.

[169] This section was originally prepared by CRS Analyst, Linda J. Schierow, now retired.

[170] J. Pettis, E. Lichtenberg, M. Andree, et al., "Crop Pollination Exposes Honey Bees to Pesticides Which Alters Their Susceptibility to the Gut Pathogen *Nosema ceranae*," *PLoS ONE*, v. 8. n. 7 (2013), p. e70182.

[171] Fungicides typically are applied to blooming plants which may be visited by bees. Labels on fungicides recommend such application for control of certain pests.

[172] Beebread is a pollen mixture stored in honeycomb cells and used with honey by bees as food. Beebread provides nutrition for the queen and brood.

[173] Trapped pollen refers to pollen collected at the entrance to the hive, removed from bees with the use of a trap.

[174] This is wax originating in the area of the hive where brood is reared.

[175] Foundation is the hexagonal sheets of wax and or wax/plastic combination that serves as a basis for comb construction by honey bees.

[176] C. A. Mullin, M. Frazier, J. L. Frazier, et al., "High Levels of Miticides and Agrochemicals in North American Apiaries: Implications for Honey Bee Health," *PLoS ONE*, v. 5, n. 3 (2010), p. e9754.

[177] C. Krupke, G. Hunt, B. Eitzer, et al., "Multiple Routes of Pesticide Exposure for Honey Bees Living near Agricultural Fields," *PLoS ONE*, v. 7, n. 1 (2012), p. e29268.

[178] G. C. Cutler and C. D. Scott-Dupree, "Exposure to Clothianidin Seed-Treated Canola Has No Long-Term Impact on Honey Bees," *Journal of Economic Entomology*, v. 100 (2007), pp. 765-772.

[179] J. M. Bonmatin, P. A. Marchand, R. Cahvet, et al., "Quantification of Imidacloprid Uptake in Maize Crops," *Journal of Agricultural and Food Chemistry*, v. 53 (2005), pp. 5336-5341; and C. Krupke, G. Hunt, B. Eitzer, et al., "Multiple Routes of Pesticide Exposure for Honey Bees Living near Agricultural Fields," *PLoS ONE*, v. 7, n. 1 (2012), p. e29268.

[180] F. M. Laurent and E. Rathahao, "Distribution of [^{14}C] Imidacloprid in Sunflowers (*Helianthus annuus L.*) Following Seed Treatment," *Journal of Agricultural and Food Chemistry*, v. 51 (2003), pp. 8005-8010; R. Schmuck, R. Schoning, A. Stork, et al., "Risk Posed to Honeybees (*Apis mellifera* L, Hymenoptera) by an Imidacloprid Seed Dressing of Sunflowers," *Pest Management Science*, v. 57 (2001), pp. 225-238.

[181] K A. Stoner and B. D. Eitzer, "Movement of Soil-Applied Imidacloprid and Thiamethoxam into Nectar and Pollen of Squash (*Cucurbita pepo*)," *PLoS ONE*, v. 7, n. 6 (2012), p. e39114.

[182] G. P. Dively and A. Kamel, "Insecticide Residues in Pollen and Nectar of a Cucurbit Crop and Their Potential Exposure to Pollinators," *Journal of Agricultural and Food Chemistry*, v. 60 (2012), pp. 4449-4456.

[183] K. A. Stoner and B. D. Eitzer, "Movement of Soil-Applied Imidacloprid and Thiamethoxam into Nectar and Pollen of Squash (*Cucurbita pepo*)," *PLoS ONE*, v. 7, n. 6 (2012), p. e39114.

[184] C. Krupke, G. Hunt, B. Eitzer, et al., "Multiple Routes of Pesticide Exposure for Honey Bees Living near Agricultural Fields," *PLoS ONE*, v. 7, n. 1 (2012), p. e29268. The study references other studies with similar findings, including H. V. V. Tomé, G. F. Martins, M. August, et al., "Imidacloprid-Induced Impairment of Mushroom Bodies and Behavior of the Native Stingless Bee *Melipona quadrifasciata anthidioides*," *PLoS ONE*, v. 7, n. 6 (2012), p. e38406; D. M. Eiri and J. C. Nieh, "A Nicotinic Cacetylcholine Receptor Agonist Affects Honey Bee Sucrose Responsiveness and Decreases Waggle Dancing," *Journal of Experimental Biology*, v. 215, n. 12 (2012), pp. 2022- 2029; A. Decourtye, E. Lacassie, and M. Pham-Delegue, "Learning Performances of Honey Bees (*Apis mellifera L.*) Are Differentially Affected by Imidacloprid According to the Season," *Pest Management Science*, v. 59 (2003), pp. 269-278; or V. Mommaerts, S. Reynders, J. Boulet, et al., "Risk Assessment for Side-Effects of Neonicotinoids Against Bumblebees with and Without Impairing Foraging Behavior," *Ecotoxicology*, v. 19 (2010), pp. 207-215.

[185] J. Pettis, E. Lichtenberg, M. Andree, et al., "Crop Pollination Exposes Honey Bees to Pesticides Which Alters Their Susceptibility to the Gut Pathogen *Nosema ceranae*, *PLoS ONE*, v. 8. n. 7 (2013), p. e70182; P. R. Whitehorn, S. O'Connor, F. L. Wackers, et al., "Neonicotinoid Pestide Reduces Bumble Bee Colony Growth and Queen Production," *Science*, v. 336 (2012), pp. 351-352; and T. Blacquiere, G. Smagghe, C. A. M. van Gestel, et al., "Neonicotinoids in Bees: A Review on Centrations, Side-Effects and Risk Assessment, *Ecootoxicology*, v. 21 (2012), pp. 973-992.

[186] EFSA press release, "EFSA Press Release: EFSA Identifies Risks to Bees from Neonicotinoids," January 16, 2013.

[187] Chensheng Lu, K. M. Warchol, and R. A. Callahan, "In Situ Replication of Honey Bee Colony Collapse Disorder," *Bulletin of Insectology*, v. 65, n. 1 (March 2012), pp. 99-106.

[188] G. DeGrandi-Hoffman, D. Sammataro, and R. Simonds, "Are Agrochemicals Present in High Fructose Corn Syrup Fed to Honey Bees (*Apis mellifera* L.)?" *Journal of Apicultural Research*, v. 51, n. 4 (2012), pp. 371-372.

[189] C. A. Johansen and D. F. Mayer, 1990, *Pollinator Protection, A Bee and Pesticide Handbook.*

[190] 2007 NRC study, p. 79.

[191] W. Mao, M. A. Schuler, and M. R. Berenbaum, "CYP9Q-Mediated Detoxification of Acaricides in the Honey Bee (*Apis mellifera*)," *Proceedings of the National Academy of Sciences of the United States of America*, v. 108, n. 31 (2011), pp. 12657-12662; and R. M. Johnson, H. S. Pollock, and M. R. Berenbaum, "Synergistic Interactions Between In-Hive Miticides in *Apis mellifera*," *Journal of Exonomic Entomology*, v. 102, n. 2 (2009), pp. 474-479.

[192] R. M. Johnson, L. Dahlgren, B. D. Siegfried, et al., "Acaricide, Fungicide and Drug Interactions in Honey Bees (*Apis mellifera*)," *PLoS ONE*, v. 8, n. 1 (2013), p. e54092.

[193] T. Blacquiere, et al., "Neonicotinoids in Bees: A Review on Concentration, Side-Effects and Risk Assessment," *Ecotoxicology*, vol. 21 (2012), pp. 973-992; and also comments by

Nathalie Steinhauer (University of Maryland) to the Virginia Native Plant Society, October 9, 2014.

[194] C. A. Mullin, M. Frazier, J. L. Frazier, et al., "High Levels of Miticides and Agrochemicals in North American Apiaries: Implications for Honey Bee Health," *PLoS ONE*, v. 5, n. 3 (2010), p. e9754.

[195] C. A. Mullin, M. Frazier, J. L. Frazier, et al., "High Levels of Miticides and Agrochemicals in North American Apiaries: Implications for Honey Bee Health," *PLoS ONE*, v. 5, n. 3 (2010), p. e9754.

[196] H. Thompson and S. Wilkins, "Assessment of the Synergy and Repellency of Pyrethroid/Fungicide Mixtures," *Bulletin of Insectology*, v. 56 (2003), pp. 131-0134; and R. L. Cox and W. T. Wilson, "Effects of Permethrin on the Behavior of Individually Tagged Honey Bees, *Apis mellifera* L. (Hymenooptera, Apidae)," *Environmental Entomology*, v. 13 (1984), pp. 375-378.

[197] Findings of the task force (which is affiliated with the International Union for Conservation of Nature) are at http://www.tfsp.info. The published study is M. B. van Lexmond, et al., "Worldwide Integrated Assessment on Systemic Pesticides," *Environ Sci Pollut Res*, August 2014. See also M. Chagnon and D. Kreutzweiser, et al., "Risks of Large-Scale Use of Systemic Insecticides to Ecosystem Functioning and Services," *Environ. Sci. Pollut. Res.*, July 2014. See also D. Gibbons, C. Morrisey, P. Mineau, "A Review of the Direct and Indirect Effects of Nenicotinoids and Fipronil on Vertebrate Wildlife," *Environ. Sci. Pollut. Res.*, June 2014.

[198] Press release, "Systemic Pesticides Pose Global Threat to Biodiversity and Ecosystem Services," June 24, 2014.

[199] D. Cox-Foster and D. vanEngelsdorp, "Solving the Mystery of the Vanishing Bees," *Scientific American*, March 31, 2009; Northwest Coalition for Alternatives to Pesticides, "Imidacloprid, Fact Sheet," *Journal of Pesticide Reform*, Spring 2001 (http://www.pesticide.org/imidacloprid.pdf); and Apiculteurs de France, "Composite Document of Present Position Relating to Gaucho, Sunflower and Bees" (http://www.beekeeping.com/articles/us/gaucho/ manifestation_paris_us.htm). See also J. Cummins, "Neonicotinoid Insecticides Used in Seed Dressing May Be Responsible for the Collapse of Honeybee Colonies," Institute of Science in Society, April 24, 2007.

[200] BNA's *International Environmental Law Committee Newsletter*, vol. 11, no. 1, February 2009.

[201] M. Henry, M. Beguin, F. Requier, et al., "A Common Pesticide Decreases Foraging Success and Survival in Honey Bees," *Science*, v. 336, n. 6079 (2012), pp. 348-350; and K. A. Stoner and B. D. Eitzer, "Movement of Soil-Applied Imidacloprid and Thiamethoxam into Nectar and Pollen of Squash (*Cucurbita pepo*)," *PLoS ONE*, v. 7, n. 6 (2012), p. e39114.

[202] C.W. Schneider, J. Tautz, B. Grünewald, et al., "RFID Tracking of Sublethal Effects of Two Neonicotinoid Insecticides on the Foraging Behavior of *Apis mellifera*." *PLoS ONE*, v. 7, n. 1 (2012), p. e30023.

[203] Bayer Crop Science, "Overview of Recent Publications on Neonicotinoids and Pollinators." Paper was prepared by Richard Heintzelman, Iain D. Kelly, David L. Fischer, and Christian Maus, May 23, 2012 (referred to here as Heintzelman et al.). Paper was submitted to CRS from Jean Reimers (CropLife America), June 20, 2012.

[204] Opinions submitted by Christian Krupke (Purdue University) and James Frazier (Pennsylvania State University) in response to a review of publications on neonicotinoids and pollinators prepared by Heintzelman et al. (May 23, 2012). These comments are contained within Appendix A and Appendix B, respectively, as part of public comments submitted to EPA by

the Center for Food Safety (docket: EPA-HQ-OPP-2012-0334). The Center for Food Safety's own review of the scientific literature is titled "Pollinators and Pesticides," 2013.

205 Ibid. Appendix B, "Opinion of Prof. James Frazier," June 14, 2012.

206 Ibid. Appendix A, "Opinion of Prof. Christian Krupke," June 7, 2012.

207 J. DeCant and M. Barrett, 2010, "Environmental Fate and Ecological Risk Assessment for the Registration of Clothianidin for Use as a Seed Treatment on Mustard Seed (Oilseed and Condiment) and Cotton," EPA/Office of Prevention, Pesticides and Toxic Substances, Office of Pesticide Programs, Washington, DC, p. 14.

208 A. Tapparo, D. Marton, C. Giorio, et al., "Assessment of the Environmental Exposure of Honeybees to Particulate Matter Containing Neonicotinoid Insecticides Coming from Corn Coated Seeds," *Environmental Science & Technology*, vol. 46, no. 5 (2012), pp. 2592–2599.

209 V. Girolami, M. Marzaro, L. Vivian, et al., "Fatal Powdering of Bees in Flight with Particulates of Neonicotinoids Seed Coating and Humidity Implication," *Journal of Applied Entomology*, v. 136, n. 1-2 (2012), pp. 17-26.

210 Health Canada, "Evaluation of Canadian Bee Mortalities that Coincided with Corn Planting in Spring 2012," 2013.

211 Judy Y. Wu, Carol M. Anelli, and Walter S. Sheppard, "Sub-Lethal Effects of Pesticide Residues in Brood Comb on Worker Honey Bee (*Apis mellifera*) Development and Longevity," *PLoS ONE*, v. 6, n. 2 (2011), p. e14720.

212 En-Cheng Yang, et al., "Impaired Olfactory Associative Behavior of Honeybee Workers Due to Contamination of Imidacloprid in the Larval Stage," *PLoS ONE*, v. 7. n. 11 (November 2012), p.e49472.

213 T. Matsumoto, "Reduction in Homing Flights in the Honey Bee *Apis mellifera* After a Sublethal Dose of Neonicotinoid Insecticides," *Bulletin of Insectology*, vol. 66, no. 1 (2013), pp. 1-9.

214 J. Y. Wu, M. D. Smart, C. M. Anelli, et al., "Honey Bees (*Apis mellifera*) Reared in Brood Combs Containing High Levels of Pesticide Residues Exhibit Increased Susceptibility to *Nosema* (Microsporidia) Infection," *Journal of Invertebrate Pathology*, v. 109 (2012), pp. 326-329; and J. Pettis, D. vanEngelsdorp, J. Johnson, et al., "Pesticide Exposure in Honey Bees Results in Increased Levels of the Gut Pathogen *Nosema*," *Naturwissenschaften*, v. 99 (2012), pp. 153-158.

215 C. Alaux, J.-L. Brunet, C. Dussaubat, et al., "Interactions Between *Nosema* Microspores and a Neonicotinoid Weaken Honeybees (*Apis mellifera)*," *Environmental Microbiology*, v. 12 (2012), pp. 774-782; J. Y. Wu, M. D. Smart, C. M. Anelli, et al., "Honey Bees (*Apis mellifera*) Reared in Brood Combs Containing High Levels of Pesticide Residues Exhibit Increased Susceptibility to *Nosema* (Microsporidia) Infection," *Journal of Invertebrate Pathology*, v. 109 (2012), pp. 326-329; J. Pettis, D. vanEngelsdorp, J. Johnson, et al., "Pesticide Exposure in Honey Bees Results in Increased Levels of the Hut Pathogen *Nosema*," *Naturwissenschaften*, v. 99 (2012), pp. 153-158; and J. Pettis, E. Lichtenberg, M. Andree, et al., "Crop Pollination Exposes Honey Bees to Pesticides Which Alters their Susceptibility to the Gut Pathogen *Nosema ceranae*, *PLoS ONE*, v. 8. n. 7 (2013), p. e70182.

216 G. Di Priscoa, V. Cavaliereb, D. Annosciac, et al., "Neonicotinoid Clothianidin Adversely Affects Insect Immunity and Promotes Replication of a Viral Pathogen in Honey Bees," *Proceedings of the National Academy of Sciences*, v. 110, no. 46 (2013), pp. 18466–18471. See also F. Sánchez-Bayo, "The Trouble with Neonicotinoids," *Science*, v. 346, issue 6211, November 14, 2014.

217 Waggle dancing refers to when a bee communicates food locations to nest mates.

[218] D. M. Eiri and J. C. Nieh, "A Nicotinic Cacetylcholine Receptor Agonist Affects Honey Bee Sucrose Responsiveness and Decreases Waggle Dancing," *Journal of Experimental Biology*, v. 215, n. 12 (2012), pp. 2022- 2029.

[219] Chensheng Lu, K. M. Warchol, and R. A. Callahan, "In Situ Replication of Honey Bee Colony Collapse Disorder," *Bulletin of Insectology*, v. 65, n. 1 (2012), pp. 99-106.

[220] H. V. V. Tomé, G. F. Martins, M. August, et al., "Imidacloprid-Induced Impairment of Mushroom Bodies and Behavior of the Native Stingless Bee *Melipona quadrifasciata anthidioides*," *PLoS ONE*, v. 7, n. 6 (2012), p. e38406. The mushroom body refers to a specific region of the bee brain.

[221] S. M. Williamson and G. A. Wright, "Exposure to Multiple Cholinergic Pesticides Impairs Olfactory Learning and Memory in Honeybees," *Journal of Experimental Biology*, v. 216, n. 10 (February 2013), pp. 1799-1807.

[222] M. J. Palmer, et al., "Cholinergic Pesticides Cause Mushroom Body Neuronal Inactivation in Honeybees," *Nature Communications*, v. 4, n. 1634 (March 2013).

[223] *Nosema* disease is a common bee disease, caused by a microscopic fungus (formerly considered to be a protozoan).

[224] J. Aufauvre, D. G. Biron, C. Vidau, et al., "Parasite-Insecticide Interactions: A Case Study of *Nosema ceranae* and Fipronil Synergy on Honeybee," *Scientific Reports* vol. 2 (2012), no. 326; C. Vidau, M, Diogon, J. Aufauvre, et al., "Exposure to Sublethal Doses of Fipronil and Thiacloprid Highly Increases Mortality of Honeybees Previously Infected by *Nosem ceranae*," *PLoS ONE*, v. 6, n. 6 (2011), p. e21550; and C. Alaux, J. Brunet, C. Dussaubat, et al., "Interactions Between *Nosema* Microspores and a Neonicotinoid Weaken Honeybees (*Apis mellifera*)," *Environmental Microbiology*, v. 12, n. 3 (2012), pp. 774-782; and J. Pettis, E. Lichtenberg, M. Andree, et al., "Crop Pollination Exposes Honey Bees to Pesticides Which Alters Their Susceptibility to the Gut Pathogen *Nosema ceranae*, *PLoS ONE*, v. 8. n. 7 (2013), p. e70182.

[225] Entombing refers to when bees seal up hive cells containing pollen to remove them from use and protect the hive from their contents. D. vanEngelsdorp, J. D. Evans, L. Donovall, et al., "'Entombed Pollen': A New Condition in Honey Bee Colonies Associated with Increased Risk of Colony Mortality," *Journal of Invertebrate Pathology*, v. 101 (2009), pp. 147–149.

[226] W. Zhu, et al., "Four Common Pesticides, Their Mixtures and a Formulation Solvent in the Hive Environment Have High Oral Toxicity to Honey Bee Larvae," *PLoS ONE*, v. 9, n. 1, p. e77547.

[227] Chensheng Lu, K. M. Warchol, and R. A. Callahan, "Sub-Lethal Exposure to Neonicotinoids Impaired Honey Bees Winterization Before Proceeding to Colony Collapse Disorder," *Bulletin of Insectology*, v. 67, n. 1 (2014), pp. 125-130.

[228] Comments attributable to Jeff Pettis (USDA) and other university researchers, and also representatives of the agrochemical industries. See, for example, A. Boyle, "Hope for the Honeybees?" *NBC News.com*, May 14, 2014; and T. Philpott, "Did Scientists Just Solve the Bee Collapse Mystery?" *Mother Jones*, May 20, 2014.

[229] For example, presentations of preliminary findings of ongoing studies by Stephen Cook (USDA ARS Beltsville Bee Lab) to DC Beekeepers Alliance, September 17, 2014; and presentations of unpublished work by Nathalie Steinhauer (University of Maryland) to the Virginia Native Plant Society, October 9, 2014.

[230] Presentations of preliminary findings of ongoing studies by Stephen Cook (USDA ARS Beltsville Bee Lab) to DC Beekeepers Alliance, September 17, 2014.

[231] C. Moffat, J.G. Pacheco, S. Sharp, et al., "Chronic Exposure to Neonicotinoids Increases Neuronal Vulnerability to Mitochondrial Dysfunction in the Bumblebee (*Bombus*

terrestris)," *Journal of the Federation of American Societies for Experimental Biology (FASEB)*, January 29, 2015.

[232] P. R. Whitehorn, S. O'Connor, D. Goulson, et al., "Neonicotinoid Pesticide Reduces Bumble Bee Colony Growth and Queen Production," *Science*, v. 336 (2012), pp. 351-352.

[233] V. Mommaerts, S. Reynders, J. Boulet, et al., "Risk Assessment for Side-Effects of Neonicotinoids Against Bumblebees with and Without Impairing Foraging Behavior," *Ecotoxicology*, v. 19 (2010), pp. 207-215.

[234] I. Laycock, K. M. Lenthall, A. T. Barratt, and J. E. Cresswell, "Effects of Imidacloprid, a Neonicotinoid Pesticide, on Reproduction in Worker Bumble Bees (*Bombus terrestris*)," *Ecotoxicology*, vol. 7 (2012), pp. 1937-1945.

[235] J. E. Cresswell, et al., "Clearance of Ingested Neonicotinoid Pesticide (Imidacloprid) in Honey Bees (*Apis mellifera*) and Bumblebees (*Bombus terrestris*)," *Pest Management Science*, v. 70, no. 2 (February 2014), pp. 332–337.

[236] See, for example, A. E. Gradish, et al., "Effect of Reduced Risk Pesticides for Use in Greenhouse Vegetable Production on *Bombus impatiens* (Hymenoptera: Apidae)," *Pest Management Science*, v. 66, no. 2 (2010), pp. 142-146.

[237] C. Elston, et al., "Sub-Lethal Effects of Thiamethoxam, a Neonicotinoid Pesticide, and Propiconazole, a DMI Fungicide, on Colony Initiation in Bumblebee (*Bombus terrestris*) Micro-Colonies," *Apidologie*, 2013.

[238] R. J. Gill, O. Ramos-Rodriguez, and N. E. Raine, "Combined Pesticide Exposure Severely Affects Individual- and Colony-Level Traits in Bees," *Nature*, v. 491, n. 7422 (2012), pp. 105-108.

[239] Bayer Crop Science, "Overview of Recent Publications on Neonicotinoids and Pollinators." Unpublished paper received from Jean Reimers (CropLife America), June 20, 2012.

[240] UK Department of Environmental Food and Rural Affairs (DEFRA), Food and Environment Research Agency, "Effects of Neonicotinoid Seed Treatments on Bumble Bee Colonies Under Field Conditions," March 2013.

[241] See, for example, J. Entine, "Bee Deaths Reversal: As Evidence Points Away from Neonics as Driver, Pressure Builds to Rethink Ban," opinion piece published in *Forbes* magazine, February 5, 2014.

[242] See, for example, comments by University of Sussex professor Dave Goulson, "One More unto the Breach; a Look at DEFRA's Stance on Neonicotinoids," September 13, 2013, http://splash.sussex.ac.uk/blog/for/dg229.

In: Bee Health
Editor: Cristina Weaver
ISBN: 978-1-63482-671-6

Chapter 3

GUIDANCE FOR ASSESSING PESTICIDE RISKS TO BEES[*†]

Office of Pesticide Programs, EPA

EXECUTIVE SUMMARY

This document provides guidance to risk assessors for evaluating the potential risk of pesticides to bees, particularly honey bees (*Apis mellifera*). This guidance is not limited to identifying the risk assessment process but includes consideration of the underlying data on which the process is based. For purposes of brevity, this guidance refers to the *White Paper in Support of the Proposed Risk Assessment Process for Bees*[1] submitted to the Federal Insecticide, Fungicide and Rodenticide Act (FIFRA) Scientific Advisory Panel (SAP) for review and comment in September 2012. The White Paper describes the basic framework of the risk assessment process and the data used to inform the various tiers of refinement that may be required to support risk management decisions. This guidance also considers recommendations[2] provided by the FIFRA SAP in response to the White Paper, where such recommendations can be immediately implemented. Additional

* This is an edited, reformatted and augmented version of a document issued by the U.S. Environmental Protection Agency, June 19, 2014. Additional contributions were made by Health Canada Pest Management Regulatory Agency and the California Department of Pesticide Regulation.

† Currently, due to resource limitations, the California Department of Pesticide Regulation does not conduct full ecological risk assessments, but reserves the right to do so in the future.

recommendations from the FIFRA SAP that cannot be implemented at this time, because the science supporting such efforts has not been sufficiently vetted, will be considered as the science evolves.

The risk assessment process described in the White Paper identifies a tiered approach where data are collected on individual bees that are representative of different life stages (larval/pupal versus adults) and castes (*e.g.*, worker bees). While additional data may be available on other bee species and these data can be included in the tiered risk assessment process as an additional line of evidence, the primary process relies on honey bee data as a surrogate for both *Apis* and non-*Apis* bees. In this process, laboratory-based studies of larval/pupal and adult honey bees provide data on individual bees that can be used as a surrogate for other species of bees, including solitary species. At the semi-field and full-field levels, studies of the colony can be used to represent effects to honey bees themselves and as a surrogate for other social bees. An advantage of using honey bees is that the husbandry and life cycle of the species and its significance in pollination services is well known and test protocols are available. As the science evolves, methods and studies using non-*Apis* bees may be considered and incorporated into the risk assessment.

The risk assessment process for bees is consistent with that used for other taxa, as described in the Overview Document[3], in that it consists of three phases (*i.e.*, problem formulation, analysis, and risk characterization) and it is tiered. The first tier consists of a screening-level risk assessment that is intended to be sufficiently conservative such that chemicals that pass the screen are considered to represent a relatively low risk of adverse effects to bees. For those chemicals which do not pass the initial screen, refinements in exposure estimates and/or mitigation measures may sufficiently reduce risk quotients (RQs) below levels of concern (LOCs) such that further refinements are not needed. For chemicals where RQ values still exceed LOCs and, depending on risk management needs, additional refinements in exposure and/or effects estimates can be made based on studies with increasing levels of environmental realism. Although approaches for estimating exposure and effects differ across aquatic and terrestrial systems as well as between plants and animals, the basic process of moving from a screening-level assessment with conservative assumptions to more refined measures of exposure and effects is consistent across taxa.

The guidance follows the generic structure of the risk assessment process, as described in the White Paper and Overview Document. While this guidance is not intended to be exhaustive, it provides staff with sufficient information with which to ensure consistency in ecological risk assessments written in support of new and existing pesticide registration decisions. The different levels of refinement described in the guidance are not intended to be prescriptive; the specific set of data used in assessing potential risks of a pesticide to bees ultimately depends on multiple lines of evidence and risk management objectives.

1. Overview of Pollinator Risk Assessment Process

This section summarizes the overall risk assessment process for characterizing the risks of pesticides to honey bees (*Apis mellifera*), which are used as a surrogate species for other *Apis* and non-*Apis* bees and other insect pollinators. It provides a brief overview of the key steps and decision points involved in the risk assessment process. As such, it should be not used in isolation; rather it should be considered in conjunction with the detailed risk assessment guidance described in the ensuing sections on Problem Formulation (**Section** 2), Analysis (**Section** 3), and Risk Characterization (**Section** 4). If acceptable data are available for non-*Apis* bees, that information should be considered as a line of evidence in determining potential risks to bees. An illustration of the decision-making process for assessing risks to honey bees from foliar spray applications of pesticides is shown in **Figure 1** while that for pesticides applied via soil or seed treatments is shown in **Figure 2**. The overall approach is a tiered process whereby risks are first assessed using simple and conservative exposure screening models to generate estimated environmental concentrations (EECs) coupled with toxicity estimates derived from laboratory studies (Tier I) to calculate risk quotients (RQs) for individual bees. Pending results of the Tier I risk estimation, consideration is given to collecting and evaluating information at higher tiers (Tiers II and III), where results are based on effects to the colony, and are considered more environmentally realistic but also requiring greater resources to conduct and interpret. A step-by-step summary of the processes is provided below. The process described below is not intended to be prescriptive and risk assessors should evaluate multiple lines of evidence in determining which data to recommend and be cognizant of the timeline for risk management decisions. Across all of the scenarios described below, an initial determination must be made as to whether a reasonable potential exists for exposure of bees to the pesticide of concern. If there is no exposure, then the likelihood of adverse effects (*i.e.*, risk) is presumed to be low and an additional assessment is not warranted. In such cases, the risk assessor should provide an explanation for why the use of the pesticide will not be likely to result in exposure of concern to bees. Also, and consistent with the process used to assess risk to other taxa, the risk assessment process is intended to be iterative, and the risk assessor should consider the effect that mitigation options may have on reducing exposure and thereby mitigating the need for additional refinements.

1.1. Foliar Spray Applications

Step 1. Determine if Bees May Be Exposed. As part of Problem Formulation (Section 2), information on the pesticide use characteristics (Box 1, Figure 1), chemical properties and potential exposure routes are evaluated to determine the need for conducting a bee risk assessment. Information provided with the generic conceptual models for bee risk assessment should be consulted (Appendix 1. Conceptual Models). In general, outdoor spray applications are assumed to have a reasonable potential to result in exposure of adult bees and their brood (eggs, larvae, pupae) to pesticides if they are applied to pollinator attractive crops or drift to pollinator attractive plants during periods when bees are likely to be foraging. Exposure to bee brood and other castes of bees in hives is expected when exposure to foraging bees is identified, as foragers will bring residues back to the hive. Pre-bloom foliar application of pesticides to pollinator attractive crops may also result in exposure to bees if the pesticide is persistent and may translocate to pollen and nectar after spray application. In contrast, indoor uses are generally assumed not to have a reasonable potential for exposure of bees to pesticides. Exceptions do occur, such as in greenhouses where bees are used for pollination (*e.g.*, bumble bee pollination of tomatoes).

Step 2. Calculate Tier I Screening-Level Risks. If a reasonable potential for exposure to the pesticide is identified, a screening-level risk assessment is conducted. This step involves a comparison of Tier I estimated exposure concentrations (EECs) for contact and oral routes of exposure to adults and larvae (Boxes 3a,b,c; Section 3.1) to Tier I acute and chronic levels of effects to individual bees using laboratory-based studies (e.g., acute median lethal dose (LD50), chronic no observable adverse effect concentration (NOAEC); Section 1.1). The Tier I EECs can be estimated using the Bee-REX model (see Appendix 3. Bee REX for additional details). The conservatism of the Tier I screening-level risk quotient (RQ) value results primarily from the model-generated exposure estimates that, while intended to represent environmentally relevant exposure levels, are nonetheless considered high-end estimates. The resulting acute and chronic RQ values (Boxes 4a,b,c) are then compared to the corresponding level of concern (LOC) values for acute and chronic risk (*i.e.*, 0.4 and 1.0, respectively). Generally, if RQ values are below their respective LOCs, a presumption of minimal risk is made, since the Tier I risk estimation methods are designed to be conservative. Risk assessors should also consider other lines of evidence in making this determination, as explained in Section 3. It is also important that any uncertainties related to screening-level estimates of exposure and/or toxicity are characterized in the risk discussion section of the risk characterization.

Step 3. Refine Tier I Screening-Level Risk Estimates. If risk concerns are identified, the Tier I assessment may be refined using additional data (Box 6; Section 3.1.2). The initial Tier I risk estimation is designed to produce conservative estimates of risk in order to minimize the occurrence of false negative findings. Refinements to Tier I risk estimates may include consideration of pesticide-specific residue data available from crop magnitude of residue studies that are relevant to bees or available studies that quantify pesticide residues in pollen and nectar (Box 9a). Refined RQ values, based on refined estimates of exposure coupled with the Tier 1 effects endpoints from individual bees, are then compared to the aforementioned LOCs to determine the potential for risk (Box 7).

Step 4. Consider Uncertainties, Risk Mitigation Options and Need for Tier II Risk Estimation. If risks are identified from Tier I, the risk assessor should consider the uncertainties associated with risk estimation, information from other lines of evidence, and the impact of any risk mitigation options identified for the pesticide of concern (Box 8). These risk mitigation options may include reductions in application rates and restriction of application methods, and recalculation of Tier I risk estimates as a result of reduced environmental loading. Restrictions on the timing of pesticide applications and crops species may also be considered to minimize exposure to bees. The risk assessor should also consider whether information on pesticide exposure and effects collected using Tier II studies are needed (e.g., residues in pollen and nectar studies described in Appendix 2. Considerations related to quantifying residues of pesticides in pollen and nectar using pesticide-specific studies; semi-field tunnel studies as described in Section 3.2.5). Tier II studies may be used to identify more targeted risk mitigation options than those that could be identified based on Tier I data. Measured residues in pollen and nectar (Box

9a) from these studies may also be used to refine risk estimates for Tier I as described previously. Tier II effect studies characterize pesticide effects at the whole-colony level and therefore, reduce uncertainty associated with extrapolating effects on individual bees under laboratory conditions (Tier I toxicity studies) to effects on the colony. It is important to recognize that Tier II effect studies are conducted under semi-field conditions where the high-end exposure at the colony level is expected. In Tier II studies other stressors may be present and potential compensatory mechanisms of the colony may occur. Tier II studies should be designed to address potential uncertainties identified in the Tier I assessment or elsewhere (*e.g.*, incident reports). Unlike Tier I, characterization of risk in Tier II does not involve the calculation of RQ values *per se* (Box 10). Rather, risks at the colony level are usually characterized in relation to pesticide application rate and/or measured residue levels (Section 3.2.6). Interpretation of such whole-colony effects studies is often much more complex than Tier I studies, and relies on comprehensive considerations of whether adverse effects are likely to occur at the colony level.

Step 5. Consider Uncertainties, Risk Mitigation Options and Need for Tier III Studies. Based on the risks identified at lower-tier assessments, their associated uncertainties, and other lines of evidence, the risk assessor should consider the impact of any risk mitigation options identified for the pesticide of concern (Box 11). The need for more refined information conducted at Tier III should be determined depending on the nature of the estimated risks, the associated uncertainties, and available risk mitigation options. Risk mitigation options may include reduced application rates, reduced application intervals, restrictions on applications at or near bloom or off-labeling use on a particular crop. As an example, effects on the ability of colonies to successfully emerge in the spring (*e.g.*, produce sufficient brood and adult bees after over-wintering) may be a concern for some pesticides/uses which are not typically addressed in lower tiers. Tier III studies are full-field studies that are designed to mimic actual pesticide applications and exposure of bees encountered in the environment (Section 3.2.6; Box 12). Tier III field studies are usually highly complex and require a high level of effort to design and conduct so as to address specific sources of uncertainties and potential risks identified in lower tiers. Because of the length and complexity of these studies, other factors affecting colony survival (*e.g.*, disease, pests, nutrition) may impact the successful completion and interpretation of these studies. As with any field study, the design and conduct of such studies is crucial to their interpretation and utility in risk assessment. Similar to risk characterization at Tier II, risk characterization at Tier III considers multiple lines of evidence available from lower Tiers and other information sources (*e.g.*, open literature) that meet the respective Agency's standard for inclusion in risk. Risk assessment conclusions are made based on the weight of evidence, available risk mitigation options, and uncertainties in the available data and methods (Section 4). At any stage of the risk assessment process, EPA/PMRA/CDPR may determine that risk mitigation is appropriate. The decision to implement risk mitigation is based on the existing analysis and does not necessarily depend on completing all three tiers of the full risk assessment process.

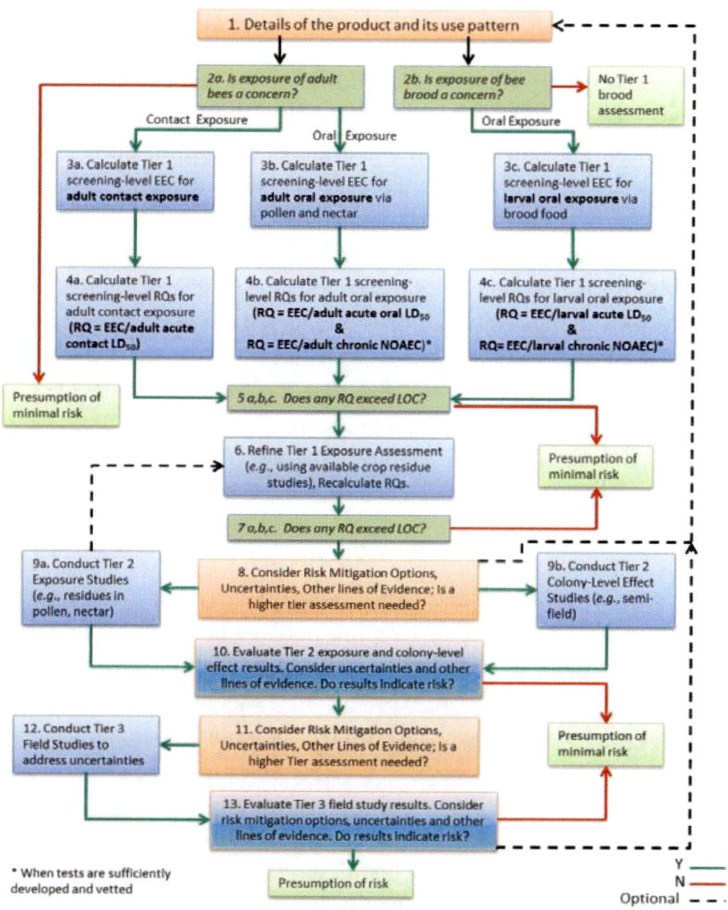

Figure 1. Tiered Approach for Assessing Risk to Honey Bees from Foliar Spray Applications.

1.2. Soil Application and Seed Treatment

The risk assessment process for evaluating soil applications (*e.g.*, soil drench) and seed treatments is similar to that described previously with foliar applications, except that risk from contact exposure is not evaluated. For soil application, it is generally assumed that exposure of honey bees from direct contact with the pesticide is minimal, given the nature of the application to bare soil, although exceptions may occur if applications are made with bee-attractive weeds present. Contact exposure of non-*Apis* bees (e.g., solitary and ground-nesting bees) may be significant with soil applications; however, the extent of this potential exposure is uncertain. It is also noted that for seed treatments, exposure of bees to pesticides has been documented via drift of abraded seed coat dust when planting under certain conditions; however, there are multiple factors determining the extent to which dust-off occurs. Modeling tools have not been developed to estimate exposure under these conditions. EPA and PMRA may determine that implementing best management practices designed to mitigate this route of exposure may be appropriate. As discussed in the preceding section, the decision to implement risk mitigation would be based on the existing analysis and does not necessarily depend on completing all three tiers of the full risk assessment process.

Step 1. Determine if Bees May Be Exposed. As part of Problem Formulation (**Section 2**), information on the pesticide use characteristics (**Box 1, Figure 2**), chemical properties and potential exposure routes is evaluated to determine the need for conducting a bee risk assessment. Information provided with the generic conceptual models for bee risk assessment should be consulted. (Appendix 1. Conceptual Models). In absence of information to indicate otherwise, it is assumed that soil-applied and seed-treated pesticides are systemic and able to be transported to pollen and nectar. Therefore, seed treatments and outdoor application to soils are generally assumed to have a reasonable potential to result in exposure of bees, including both adult and immature stages of bees, to pesticides via consumption of contaminated pollen and/or nectar.

Step 2. Calculate Tier I Screening-Level Risks. If a reasonable potential for exposure to the pesticide is identified, a screening-level risk assessment is conducted. This step is identical to that described in Step 2 for assessment of foliar spray applications except that the exposure modeling tools differ. Different methods are used to estimate pesticide residues in pollen and nectar from soil application and seed treatment (see **Section 3.1.1**). Otherwise, risk estimation, LOCs and consideration of multiple lines of evidence are identical to those described previously in Step 2 for foliar pesticide applications. All subsequent steps (*i.e.*, **Step 3. Refine Tier I Screening-Level Risk Estimates; Step 4. Consider Uncertainties, Risk Mitigation Options and Need for Tier II Refinements and Step 5. Consider Uncertainties, Risk Mitigation Options and Need for Tier III Studies**) are identical to those described previously for foliar spray application of pesticides.

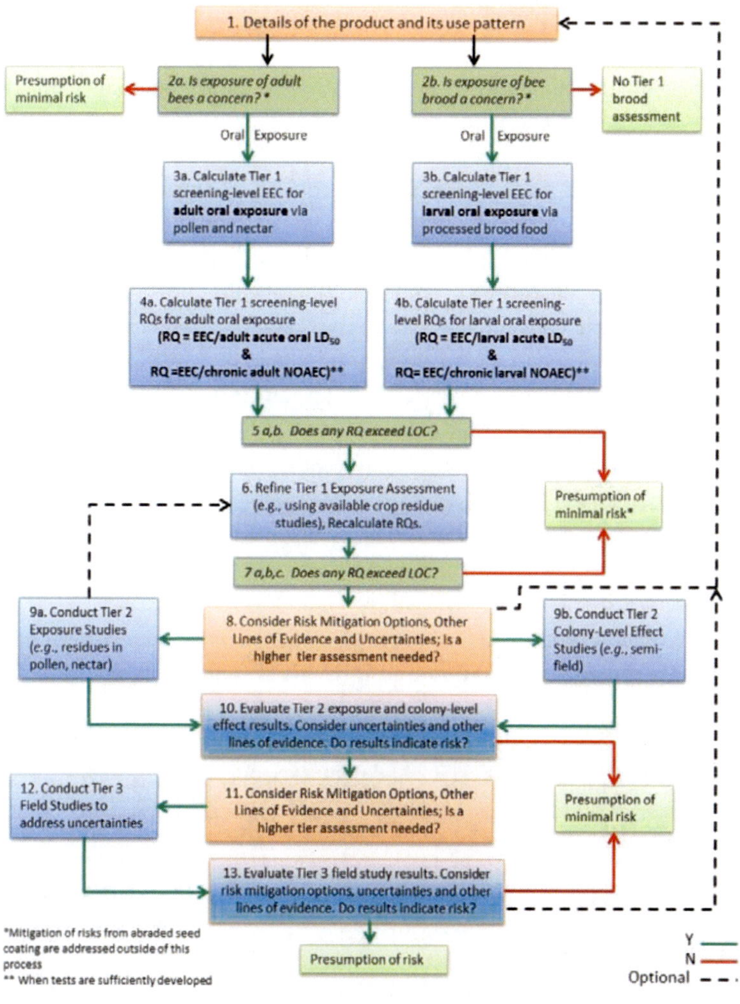

Figure 2. Tiered Approach for Assessing Risk to Honey Bees from Soil/Seed Treatments.

2. PROBLEM FORMULATION

Problem formulation is a critical step in ecological risk assessment and is intended to articulate, among other things, the protection goals around which the assessment is conducted. Relative to bees, the protection goals include the maintenance of pollination services, hive product production and biodiversity (**Table 1**). These goals do not apply uniformly across *Apis* and non-*Apis* bees; however, they are considered protective for social and solitary bees, and honey bees are generally used a surrogate for non-*Apis* bees. These protection/management goals in turn dictate assessment endpoints for which specific measurement endpoints are identified. Ideally, problem formulations should articulate the protection/management goals as well as the risk hypothesis including how these goals may be compromised due to the proposed or existing use(s) of a pesticide. The risk hypothesis and conceptual model are used to depict the hypothesis in terms of the source of the stress, route of exposure, receptor, and changes in the receptor attribute(s) of concern. A number of generic conceptual models have been developed and these should be adapted to reflect, where appropriate, the potential risks to bees that will be evaluated. The White Paper discusses and **Appendix 1.** Conceptual Models of this guidance provides conceptual models (*e.g.*, foliar application of non-systemic and systemic pesticides, soil-applied and seed-coated systemic[4] pesticides) for honey bees that can be readily adapted and integrated into conceptual models that include other taxa or included as a stand-alone conceptual model.

Table 1. Protection goals and examples of associated assessment and measurement (population and individual) endpoints for bees

Protection Goal	Assessment Endpoints	Example Measurement Endpoints	
		Population level and higher	Individual Level
Contribution to Bee Biodiversity	Species richness[1] and abundance	Individual bee survival (solitary bees) and colony strength and survival (social bees) Species richness and abundance[1]	Individual worker and larval survival assays; larval emergence; queen fecundity/reproduction

Protection Goal	Assessment Endpoints	Example Measurement Endpoints	
		Population level and higher	Individual Level
Provision of Pollination Services	Population size[2] and stability of native bees and commercially managed bees	Colony strength and survival; colony development	Individual worker and larval survival assays; queen fecundity; brood success; worker bee longevity
Production of Hive Products	Quantity and quality of hive products	Quantity and quality of hive products; including pesticide residue levels on honey/wax	Individual worker and larval survival assays; queen fecundity/reproduction; larval emergence

[1]Use of honey bees as a surrogate for other insect pollinators has limitations; however, it is assumed that as with all surrogates, data on individual organisms as well as colony-level data would provide some relevant information on the potential effects of a pesticide on both solitary bees as well as "eusocial" taxa. In addition, protection of honey bees would contribute to pollinator diversity indirectly by preserving the pollination and propagation of the many plants species pollinated by honey bees, which also serve as food sources for other pollinating insects.

[2] For managed honey bees, population size can include numbers of colonies.

For most pesticides used in an agricultural setting, the predominant exposure routes are through diet (*i.e.*, consumption of nectar and pollen) and contact (*i.e.*, direct spray). Exposure due to the vapor phase of a pesticide is relatively small compared to diet and contact, with the exception of fumigants. Additionally, the importance of exposure through consumption of drinking water, relative to the dietary and contact routes of exposure is under investigation[5]. For pesticides that are applied to seeds, exposure to dust from treated seed during planting may also be of concern. The extent to which honey bees are exposed via contact with abraded seed coat dust is determined by many factors including the physical-chemical properties of the seed coating, seed planting equipment, use of seed delivery agents (*e.g.*, talc or graphite), environmental conditions (wind speed, humidity), existence of flowers nearby the sowing area, and hive location in relation to sowing. As recognized by the FIFRA SAP, the production of dust during planting should be minimized to the extent possible in order to minimize the exposure to bees; additional considerations and assessment of risk should be completed on a chemical-specific basis.

The primary question risk assessors must address initially is whether, given the existing or proposed use(s) and physical-chemical properties of a pesticide, exposure is likely to occur for bees from the treated crop or resulting spray/dust drift to a blooming weed on the field or blooming plants near-field. If exposure is not likely for bees, then the problem formulation should state and the conceptual model should depict why the exposure is considered unlikely. Where exposure cannot be precluded, the problem formulation should identify the available body of information on exposure and effects that will be considered to support the screening-level risk assessment. If, during the risk assessment, further refinements are deemed necessary, the risk hypothesis and conceptual model can be modified accordingly and the analysis plan revised to identify additional exposure and effects data that will be considered thereby reflecting the iterative nature of the risk assessment process.

Use of honey bees as a surrogate for other insect pollinators has limitations; however, it is assumed that, as with all surrogates, data on individual organisms as well as colony-level data would provide some relevant information on the potential effects of a pesticide on both solitary bees as well as social bees. In addition, protection of honey bees would contribute to pollinator diversity indirectly by preserving the pollination and propagation of the many plants species pollinated by honey bees, which also serve as food sources for other pollinating insects. In evaluating potential risks specific to honey bees, the most important commercial pollinators, the protection goals of preserving pollination services and production of hive products (*e.g.*, honey, wax) are readily assessed through the assessment of population size and the stability (*e.g.*, presence of a queen, uniform brood pattern) of the colony and through direct and indirect measures of the quantity and quality of hive products (**Table 1**). As such, the sensitivity of individual larval or adult honey bees based on laboratory-based acute and chronic toxicity studies serve as reasonable measurement endpoints for screening-level assessments of potential adverse effects on colony strength, survival and capacity of the colony to produce any products. While these measurement and assessment endpoints are tested using managed honey bee colonies, they apply to feral honey bee colonies and, in the absence of data specific to other bees, these measurement endpoints provide useful information for assessing the survival and development of solitary bees and potential effects on bee species richness and biodiversity. To the extent that data are available for other species such as the bumble bee (*e.g.*, *Bombus terrestris*), blue orchard bee (*Osmia lignaria*), alfalfa leafcutting bee (*Megachile rotundata*), these species may also be considered in the risk assessment. As discussed in the White Paper, available information for the bumble bee, blue orchard bee and the alfalfa leafcutting bee indicate that the screening-level risk assessment based on effects to individual honey bees is likely to be protective for these three species; however, this is likely chemical specific and available data must be considered in determining whether this assumption is supported for chemical under consideration.

3. Analysis Phase

The analysis phase consists of the exposure characterization and the effects characterization relative to bees. A tiered approach, from the most conservative at lower tiers (Tier I) to more realistic at higher tiers (Tiers II and III) should be considered during the data requirement determinations and risk assessment. Steps that should be considered within each characterization are depicted in the relevant decision tree in **Figure 1 and** Figure 2.

3.1. Exposure Characterization

3.1.1. Tier I Exposure Estimates

Contact and dietary exposure are estimated separately using different approaches specific for different application methods. **Table 2** summarizes the methods used for deriving Tier I estimated environmental concentrations (EECs) for contact and dietary routes of exposure for foliar, soil, seed treatments and tree-trunk injections. These EECs are calculated using the Bee-REX model.

In the Tier I, pesticide exposures are estimated based on honey bee castes with known high-end consumption rates. For larvae, food consumption rates are based on 5-day old larvae, which consume the most food compared to other days of this life stage. For adults, the screening method relies upon nectar foraging bees, which consume the greatest amount of food (pollen and nectar) compared to other adult worker bees. It is assumed that this value will be comparable to the consumption rates of adult drones and will be protective for adult queens as well. Although the queen consumes more food than adult workers and drones, the queen consumes food that is assumed, based on currently available data, to contain orders of magnitude less pesticide than that consumed by workers since the queen is only fed a processed food, *i.e.*, royal jelly. As described in the White Paper, nectar is the major food source for foraging honey bees as well as nurse bees. Therefore, pesticide residues in nectar likely account for most of the exposures to bees, and may represent most of the potential risk concerns for adult bees. However, if residues in pollen are of concern, exposures to nurse bees, which consume more pollen than any other adult honey bees, should be considered. This is the case especially when pesticide concentrations in pollen are much greater than in nectar, or for crops that mainly provide pollen to bees and would be assessed on a case-by-case basis. In fact, the Bee-Rex allows calculation of RQs for all types of bee castes.

For chemicals with no empirical data to represent the concentration of the chemical in pollen and nectar, dietary exposure for Tier I risk assessment is estimated using generic residue data generated from other chemicals as well as other plant parts. For foliar applications for dietary exposure, it is assumed that pesticide residues on tall grass (from the Kenaga nomogram of T-REX which is incorporated into BeeREX) are a suitable surrogate for residues in pollen and nectar of flowers that are directly sprayed. For soil applications, pesticide concentrations in pollen and nectar are assumed to be consistent with chemical concentrations in the xylem of barley (calculated using the Briggs' model). For seed treatments, pesticide concentrations in pollen and nectar are based on concentrations in leaves and stems of treated plants (based on the European and Mediterranean Plant Protection Organization (EPPO) default value discussed in the White Paper), assumed to be 1 milligram per kilogram (mg/kg) or 1 part per million (ppm). More details on these methods are available in the White Paper and in the T-REX user's guidance. For tree trunk injections, methods are still under development as discussed in the White Paper; one approach is where application rates are converted into a tree-foliage weight basis (**Table 2**).

The Tier I method is intended to generate "reasonably conservative" estimates of pesticide exposure to honey bees, where reliable residue values (i.e., measured residue levels in pollen and/or nectar) are not available. As noted in **Table 2** (far right column), exposure estimates for foliar applications are derived using the application rate (AR) being assessed. Model users should follow the input parameter guidance contained in the Bee-REX user's guide when running this model.

In the Tier I approach, it is assumed that all chemicals applied as a soil drench, seed treatment, or trunk injection may be systemically transported. This assumption may be refuted using data such as Log Kow (Ryan et al. 1988) and monitoring data (e.g., crop rotation studies). Whether a chemical is transported systemically in plants could potentially be confirmed using empirical data submitted to EPA, PMRA and CADPR (e.g., plant metabolism studies); however, it would be up to pesticide registrants/applicants to submit sufficient data to demonstrate that a pesticide is not systemic.

Table 2. Summary of contact and dietary exposure estimates used for foliar applications, soil treatments, seed treatments and tree trunk injections of pesticides for Tier I risk assessments

Measurement Endpoint	Exposure Route	Exposure Estimate*
Foliar Applications		
Individual Survival (adults)	Contact	AREnglish*(2.7 μg a.i./bee) ARMetric*(2.4 μg a.i./bee)
Individual Survival (adults)	Diet	AREnglish *(110 μg a.i /g)*(0.292 g/day) ARMetric *(98 μg a.i /g)*(0.292 g/day)
Brood size and success	Diet	AREnglish *(110 μg a.i /g)*(0.124 g/day) ARMetric *(98 μg a.i /g)*(0.124 g/day)
Soil Treatments		
Individual Survival (adults)	Diet	(Briggs EEC)*(0.292 g/day)
Brood size and success	Diet	(Briggs EEC)*(0.124 g/day)
Seed Treatments		
Individual Survival (adults)	Diet	(1 μg a.i /g)*(0.292 g/day)
Brood size and success	Diet	(1 μg a.i /g)*(0.124 g/day)
Tree Trunk Applications[++]		
Individual Survival (adults)	Diet	(μg a.i. applied to tree/g of foliage)*(0.292 g/day)
Brood size and success	Diet	(μg a.i. applied to tree/g of foliage)*(0.124 g/day)

AREnglish = application rate in lbs a.i./A; ARMetric = application rate in kg a.i./ha

[*]Based on food consumption rates for larvae (0.124 g/day) and adult (0.292 g/day) worker bees and concentration in pollen and nectar. [++]Note that concentration estimates for tree applications are specific to the type and age of the crop to which the chemical is applied.

3.1.2. Refinements of Tier I Exposure

In cases where RQs exceed the LOC (discussed below), estimates of exposure may be refined using measured pesticide concentrations in pollen and nectar of treated crops, and further calculated for other castes of bees using their food consumption rates (see **Table 3**).

As discussed above in **Section 3.1.1**, the most conservative (highest) exposure estimates for contact and/or diet exposure routes are selected for the

Tier I screening-level assessment. These exposure estimates are based on adult and larval bees with the highest food consumption rates among bees. The Bee-REX tool also calculates dietary exposure values and associated RQs for larvae of different ages, adult workers with different tasks (and associated energetic requirements) and the queen. This is accomplished using the food consumption rates provided in **Table 3**. Those food consumption rates are based on work described in the White Paper (USEPA, PMRA, CDPR 2012)[6] and updated to reflect comments from the Scientific Advisory Panel (SAP). Exposure values for other groups of bees within a hive along with their RQs can be used to characterize risks of dietary exposures of different bees within the hive.

Empirical data may be used to refine conservative exposure estimates and reduce uncertainties associated with the Tier I exposure assessment by providing direct measurements of pesticide concentrations resulting from actual use settings. Studies investigating pesticide concentrations in pollen and nectar should be designed to provide residue data for crops and application methods of concern. **Appendix 2.** Considerations related to quantifying residues of pesticides in pollen and nectar using pesticide-specific studies includes considerations related to quantifying residues of pesticides in pollen and nectar using pesticide-specific studies.

Table 3. Estimated food consumption rates of bees

Life Stage	Caste (task in hive[a])	Average age (in days)[a]	Daily consumption rate (mg/day)			
			Jelly	Nectar[b]	Pollen	Total
Larval	Worker	1	1.9	0	0	1.9
		2	9.4	0	0	9.4
		3	19	0	0	19
		4	0	60[c]	1.8[d]	62
		5	0	120[c]	3.6[d]	124
	Drone	6+	0	130	3.6	134
	Queen	1	1.9	0	0	1.9
		2	9.4	0	0	9.4
		3	23	0	0	23
		4+	141	0	0	141
Adult	Worker (cell cleaning and capping)	0-10	0	60[f]	1.3 - 12[g,h]	61 - 72
	Worker (brood and queen tending, nurse bees)	6-17	0	113 - 167[f]	1.3 - 12[g,h]	114 - 179

Life Stage	Caste (task in hive[a])	Average age (in days)[a]	Daily consumption rate (mg/day)			
			Jelly	Nectar[b]	Pollen	Total
	Worker (comb building, cleaning and food handling)	11-18	0	60[f]	1.7g	62
	Worker (foraging for pollen)	>18	0	35 - 52[f]	0.041g	35 - 52
	Worker (foraging for nectar)	>18	0	292 (median)[c]	0.041g	292
	Worker (maintenance of hive in winter)	0-90	0	29f	2g	31
	Drone	>10	0	133 - 337[c]	0.0002[c]	133 - 337
	Queen (laying 1500 eggs/day)	Entire lifestage	525	0	0	525

[a] Winston (1987)

[b] Consumption of honey is converted to nectar-equivalents using sugar contents of honey and nectar.

[c] Calculated as described in this paper.

[d] Simpson (1955) and Babendreier et al. (2004)

[e] Pollen consumption rates for drone larvae are unknown. Pollen consumption rates for worker larvae are used as a surrogate.

[f] Based on sugar consumption rates of Rortais et al. (2005). Assumes that average sugar content of nectar is 30%.

[g] Crailsheim et al. (1992, 1993)

[h]Pain and Maugenet 1966

3.2. Effects Characterization

3.2.1. USEPA Toxicity Testing Requirements for Bees

USEPA data requirements for pollinator testing are currently specified in Title 40 (Protection of the Environment) of the Code of Federal Regulations, Part 158 (Data Requirements for Pesticides) Subpart G (Ecological Effects) § 158.630 (Terrestrial and Aquatic Non-target Organism Data Requirements Table).[7] When certain pesticide use patterns or triggers are met, current test requirements include the honey bee acute contact toxicity test (OCSPP Guideline 850.3020)[8], the honey bee toxicity of residues on foliage test (OCSPP Guideline 850.3030)[9] and field testing for pollinators (OCSPP Guideline 850.3040)[10]. The honey bee acute contact toxicity test is required for pesticide technical grade active ingredients (TGAI) with terrestrial, forestry and residential outdoor uses and is conditionally required for pesticides with aquatic uses as a Tier I screen conducted under laboratory conditions. If the results of the honey bee acute contact toxicity test indicates that a pesticide has a median acute lethal dose to 50% of the animals tested, *i.e.*, the LD_{50} value, of less than (<) 11 micrograms (μg) per bee and the use pattern indicates that honey bees may be exposed, then the toxicity of residues on foliage test is required as a laboratory-based test using the technical end-use product (TEP). As specified in CFR40 § 158.630[11], field testing of pollinators is required if any of the following conditions are met:

Data from other sources (Experimental Use Permit program, university research, registrant submittals, *etc.*) indicate potential adverse effects on colonies, especially effects other than acute mortality (reproductive, behavioral, *etc.*); data from residual toxicity studies indicate extended residual toxicity; or data derived from studies with terrestrial arthropods other than bees indicate potential chronic, reproductive or behavioral effects.

Field studies are intended to represent real world conditions and are considered refined (Tier III) toxicity tests. Pollinator field study designs have in the past varied considerably; therefore, study design elements that are consistent with the specific hypothesis being testing in the field study should be identified in advance and considered in the development of the study protocol. **Appendix 4.** Tier 3 Field Study Design Considerations includes some generic study design elements.

3.2.2. Additional Guidance for USEPA Pollinator Testing

In addition to the honey bee adult contact toxicity study identified in the 40CFR158, several additional studies are recommended to support the Tier I screen, as indicated in **Figure 1** and **Figure 2** and as discussed in the White Paper. These studies include:

- Acute oral toxicity to adult honey bees
- Acute oral toxicity to larval honey bees
- Chronic oral toxicity to adult honey bees
- Chronic oral toxicity to larval honey bees.

Although EPA guidelines have not been developed for these studies, the Organization for Economic Cooperation and Development (OECD) has developed a formal test guideline for an acute oral toxicity study with adult honey bees (OECD 213)[12] as well as a test guideline for acute oral toxicity study with honey bee larvae (OECD 237)[13]. As noted in the White Paper, OECD also has developed a test guideline for assessing acute contact toxicity with adult bees (OECD 214)[14], that may provide sufficient data to fulfill the 40CFR158 adult contact toxicity test requirement ordinarily fulfilled by EPA 850.3020. A chronic (repeat dose) study guidance for assessing the oral toxicity to honey bee larvae is currently in development by OECD, as is a guideline for assessing chronic (10-day) oral toxicity to honey bee adults. In situations where formal guidelines do not exist, data recommendations are identified to address specific uncertainties regarding the potential for adverse effects; these studies should be recommended as "special studies" and reference the appropriate OECD test guideline and/or other applicable guidance. Typically, non-guideline studies recommended as "special studies" (*e.g.*, acute adult oral and larval toxicity tests) are intended to address specific uncertainties and a suitable rational for the study should be included in the data recommendation.

Semi-field (*e.g.*, tunnel studies) and full-field studies have frequently been categorized as field pollinator studies falling under EPA 850.3040. As discussed in the White Paper, OECD has developed guidance for semi-field tests (*e.g.*, OECD 75)[15] and efforts are underway to further refine and standardized methods for conducting such tests. As discussed previously, Tier III studies of free-foraging bees are typically required to address specific uncertainties identified in lower tier studies and/or based on a concordance of multiple lines of evidence. Therefore, the methods used in these studies while described generically in EPA 850.3040 would likely be chemical/use specific and guidance is provided (see **Appendix 4.** Tier 3 Field Study Design Considerations) on elements to consider when recommending/evaluating such studies.

The decision to recommend bee toxicity data should consider multiple factors including the pesticide use pattern, chemical properties, and the nature of uncertainties based on existing data for the chemical or similarly structured chemicals. For example, the Tier I toxicity studies are generally limited to outdoor use patterns, although some indoor uses in greenhouses may also warrant testing when plants are commercially pollinated (e.g., bumble bee pollination of tomatoes). As discussed in the White Paper and in this guidance, if exposure is not considered likely, then effects testing may not be warranted beyond the screening-level data collected at Tier 1 or may not be warranted at all if no exposure is considered likely. If a particular use is determined not to result in exposure to bees, then a suitable rationale should be developed to support this determination. However, toxicity data using the typical end use product (TEP) may be needed in addition to data on technical grade active ingredient (TGAI) if there are data to indicate that a typical end-use product is potentially more toxic than the technical grade active ingredient and bees may come directly in contact with the intact TEP.

The need for additional data should be informed by whatever information has already been identified and should consider multiple lines of evidence. This may include data that may be available for similarly structured chemicals with common modes of action to achieve an economy of effort and limit, to the extent possible, the need for animal testing that may ultimately prove to be redundant.

While this guidance articulates a tiered progression of exposure and effects data, the agencies retain the right to ask for data in the order that is most relevant to risk management needs considering the multiple lines of evidence that are available.

3.2.3. PMRA Toxicity Testing Requirements for Bees

The PMRA has been requiring both acute oral and contact honey bee adult toxicity studies when there is potential for exposure for insect pollinators. Colony-level bee studies have also been required when there is a potential for colony exposure and effects on the colony and developing brood. These acute studies, together with colony-level studies, including Tier II semi-field and Tier III field studies, are incorporated in the risk assessment for insect pollinators. Based on the SAP Pollinator Risk Assessment Framework, additional studies are recommended to support the evaluation of insect pollinators where exposure is considered likely. These include both larval bee toxicity and adult bee chronic oral toxicity studies. Insect pollinator studies are generally not required for use site categories where pollinator exposure is not considered likely. In general, Tier I studies should be conducted under controlled laboratory conditions using the TGAI, while studies with the end-use products may be used to supplement the TGAI information. Additionally, information on potentially toxic transformation products may be required. Tier II and III studies should be conducted under progressively more realistic conditions using typical end-use products. A tiered approach is used for these data requirements; when a risk is identified at a lower tier, higher-tier studies are required to further assess the risk. Additionally, if other available information, such as scientific research and published literature, indicates potential for adverse effects on insect pollinators and/or colonies, then higher-tier studies or specific studies may be required to address the potential for concern.

Data submitted to PMRA for other non-target arthropods (predators and parasitoids) may also be considered in risk assessment for insect pollinators.

Generally, the following studies may be required to assess the risk to insect pollinators:

- Tier I insect pollinator studies:
  - Bee adult acute contact toxicity
  - Bee adult acute oral toxicity
  - Bee larvae toxicity
  - Bee adult chronic oral toxicity
- Higher Tier and other insect pollinator studies:
  - Residue study for insect pollinators
  - Semi-field study for insect pollinators
  - Field study for insect pollinators
  - Bee toxicity of residues on foliage
  - Other insect pollinator studies

3.2.4. Tier I Effects Characterization

3.2.4.1. Acute Oral and Contact Studies

Acute toxicity testing with both adult and larval bees examines the short-term effect of the test material after a short-term exposure; chronic toxicity testing typically examines multiple exposures (repeat or continuous dosing) with an extended period of observation time. Acute and/or chronic oral and contact toxicity studies should be conducted according to available study guidelines described in the previous sections. In cases where established guidelines are not available either in North America or through the OECD, proposed study protocols should be reviewed and approved prior to the initiation of the definitive study. Ideally, these acceptable protocols should be made available for consideration by other risk assessors attempting to address similar uncertainties.

Typically, the primary measurement endpoint derived from the acute oral and acute contact toxicity studies is the median lethal dose for 50% of the organisms tested (*i.e.*, LD_{50}). *A. mellifera* is used as a surrogate for assessing risks to bees. However, depending on the circumstances triggering the data requirements, risk assessors should be flexible in determining which species and studies are most relevant.

Any biological effects and abnormal responses, including sublethal effects, other than the mortality should be reported during the acute contact and oral studies according to the requirement of the study guidelines. To the extent that these measurement endpoints are dose responsive, consideration should be given to developing a median effect dose for 50% of the organisms tested (*i.e.*, ED_{50}) in addition to a median lethal dose (LD_{50}), if the data permit such an analysis. Alternatively, the risk assessment should note where sublethal measurement endpoints appear to exhibit a trend across treatment levels.

Although it may be possible to extract both a no-observed adverse effect level (NOAEL) and an ED_x or LD_x value from the laboratory-based acute studies, it is important to note differences in hypothesis-based test designs needed to reliably support NOAEL versus regression-based study designs needed to support ED_x and LD_x estimates. Hypothesis-based endpoints require sufficient replication to test for treatment effects; whereas, regression-based estimates do not. The dose-response relationship derived from Tier I studies can provide useful information in terms of slope of the dose-response curve.

3.2.4.2. Chronic Toxicity Studies

Chronic effects on adult and/or larval bees are considered in Tier I of the new assessment framework. Such effects observed in guideline toxicity tests when protocols are sufficiently vetted and/or effects that have been reported in the open literature that could potentially affect the colony should be considered in determining which, if any, higher-tier tests are required and how the study should be designed. Studies under development include the adult 10-day toxicity test and the repeat-dose larval toxicity test. Appendix O of the European Food Safety Authority (EFSA) guidance document on bee risk assessment (EFSA 2013)[16] discusses study design elements for the 10-day adult toxicity test and the OECD is currently evaluating a proposed guidance for repeat-dose testing with honeybee larvae that extends through emergence of adult bees. For the 10-day adult study, EFSA recommends the derivation of an LC50 and NOAEC while the draft OECD repeat-dose larval toxicity study recommends the derivation of an NOAEC; however, as discussed above, consideration should be given to the study design in determining the most appropriate endpoint to derive.

3.2.4.3. Toxicity Studies Involving Exposure to Vapor Phase Pesticides

For a limited number of compounds (*e.g.*, fumigants), inhalation toxicity studies may be required as a special study. Since current acute toxicity testing with young adult bees is typically conducted using a small cage where bees are exposed through the diet (spiked sugar solution), it would be possible to conduct an inhalation study using a similar test design where bees are fed untreated sugar solution, but the cage is contained within a hermetic or flow-through container containing differing concentrations of the pesticide in air. Because this is a modification of the existing protocol, registrants should submit the proposed study protocol for review and approval by EPA and/or PMRA prior to conduct of the study. The protocol should contain a rationale for the proposed air concentrations and durations of exposure.

3.2.4.4. Toxicity of Residues on Foliage to Honey Bees

Based on the contact $_{LD50}$ value, the pesticide is classified as practically non-toxic $_{(LD50}$ ≥ 11 μg/bee), moderately toxic (10.9 > LD50 >2 μg/bee), or highly toxic (<2 μg/bee). Unless the pesticide is determined to be practically non-toxic, EPA would then typically require a study on the toxicity of residues on foliage to honey bees (OSCPP 850.3030; USEPA 2012)[17].

The current OCSPP guideline study 850.3030 (USEPA 2012) evaluates the toxicity of residues on foliage to honey bees. In this study, a formulated product of a chemical is applied to a bee attractive plant (*e.g.*, clover or alfalfa) at the maximum application rate with the minimum application interval. The crop is then harvested in a manner that provides different age residues (e.g., 0, 3, 8, 24, 48, 72 and 96 hours after application). The harvested foliage is brought back to the laboratory where it is placed into cages. Adult bees are placed in each cage with the foliage and allowed to come into contact with the foliage. Food and water are provided *ad libitum*. The bees exposed to weathered foliage are monitored for a period of time until mortality declines to below 25%. The measurement endpoint derived from this study is the residual time to 25% mortality (RT25), which is defined as the time needed to reduce the "residual toxicity" of the test substance, as measured by mortality, to 25% of the test organisms. The time period determined by this toxicity value is considered to be the time that the test substance is expected to remain toxic to bees in the field from the residual contact exposure of the test substance on vegetation at an expressed rate of application (typically expressed in terms of lbs a.i./A). The results of the foliar residue study may be useful in the characterization of effects and risks of a pesticide to honey bees through a contact exposure route with treated foliage and can be used to inform risk mitigation measures related to timing of pesticide application (e.g., how long a pesticide may be expected to present a risk from contact with foliage to foraging honey bees after its application). Currently, the residue on foliage study is not used to derive RQs as part of the Tier I risk assessment process; however, the test may provide useful information for characterizing the extent (duration of time) to which residues may remain toxic to bees and may be helpful in identifying particular formulations with extended residual toxicity. It is important to note that RT25 values are specific to formulations and may not be predictive across multiple formulations of a particular active ingredient Although it may be inferred from the White Paper that an RT25 >8 hrs represents a chemical with an extended residual toxicity (ERT), ERT has not been formally defined by EPA.

3.2.5. Tier II Effects Characterization

As discussed in the White Paper, the decision to move to more refined effects testing and to transition from laboratory-based studies with individual bees (*i.e.*, Tier I) to colony-based studies (*i.e.*, Tier II or Tier III) depends on whether Tier I LOCs are exceeded, the availability of data, and the nature of uncertainties that warrant further testing. The Tier II studies are typically considered "semi-field" studies where small colonies (referred to as nuclei colonies or "nucs")[18] are enclosed in tunnels, along with pesticide-treated crops. Tier II studies may also include feeding studies in which whole colonies are tested; however, the colonies are not confined to enclosures. Typical semi-field studies are usually conducted under conditions that represent the worst-case exposure scenario of proposed uses to the entire colony (over the duration of the study) or designed to address specific uncertainties with respect to effects on the colony; whereas, the feeding studies are usually conducted with diets spiked with known concentration of test chemical using colonies that are not confined to enclosures (*i.e.*, free-foraging bees).

The Tier II study designs may be amenable to additional treatment levels and replication, thus facilitating the quantification of an application rate-response (tunnel study) or dose-response (feeding study) relationship at the colony level and determination of a NOAEC. This information may be particularly useful and transferable from the test crop/concentration to other crops where residue concentrations in pollen and nectar are available in conjunction with associated application rates.

3.2.5.1. Tunnel Design Studies

Semi-field tunnel studies provide a means of recording a number of measurement endpoints, *e.g.*, adult and larval survival, larval/pupal development, queen fecundity, worker bee behavior, to estimate various attributes regarding the whole colony. The design of a Tier II tunnel study should be flexible in addressing the specific risks identified at the Tier I risk assessment for either foliar application, seed or soil treatments, or other exposure routes of concerns with reasonable modification. Although a general study guidance for conducting a semi-field study is still under development, the OECD 75 guidance document on honey bee brood testing[19], and the European and Mediterranean Plant Protection Organization (EPPO) 170 describe basic semi-field study elements that should be considered.

In a tunnel study there is typically a pesticide exposure period in the tunnel and an extended observation period when test bees are allowed to freely forage from the landscape. While typically colonies can only be maintained in enclosures for a limited exposure time (~10 days), these colonies may be monitored following their removal from the enclosure to evaluate extended effects resulting from the exposure period or delayed exposure from ingestion of stored pollen/nectar. Each of these phases of the study presents a number of concerns. Some of the most important considerations include the following topics:

Bees in the tunnel: Food resource: It is important to ensure to the extent possible that bees have sufficient forage and that the source of pollen and nectar is not depleted during the study. In tunnel studies, small hives, referred to as "Nucs", are used so as not to overwhelm the foraging capacity contained within the enclosure. Because of the limited bloom period for most plants used in enclosure studies and because of the stress on bees confined to a limited foraging area, the exposure duration of these studies is usually about 10 days, including acclimation period.

Bees in the tunnel: Stress: Bees are intended to fly freely, and confinement to a tunnel may cause stress for the bees. Reducing confinement stress caused by the tunnels is another reason the duration of tunnel studies are generally limited to 10 days.

Behavior of test bees: Honey bee foragers are intended to fly freely in a large area. Limited tent space may have an impact on the behavior of those bees. Therefore, caution should be taken in the interpretation the results that are related to the behavior of the test bees.

Acclimation period: Long acclimation periods (*i.e.*, >3 days) within the tunnel should be avoided as they can further limit the exposure period; however, if acclimation periods are not sufficient, bees may not begin foraging efficiently until after residues on treated plants have started to decline. Therefore, initial mortality counts for bees that have not been adequately acclimated to the enclosure may not be reliable estimates of treatment-related effects. On the other hand, for seed treatment and/or soil application of a systemic pesticide, residues in pollen and/or nectar are expected at some point in time post application. In these cases, information on application timing and maximum resulting residues are necessary to ensure appropriate exposures (e.g., peak concentrations) are included during the in-tunnel phase of the study. Such exposure considerations may negate the ability to include a pesticide-free acclimation period in the tunnels.

Climate and Success: Since these studies are conducted outside, initiation of the study may have to be delayed due to inclement weather where it may not be feasible to move ahead with applications of the test material.

Free Foraging Site Location: Depending on the objective of study where an extended observation period may be required, the study report must also describe potential forage sites for the colonies and once the hives are removed from the tunnel enclosures and allowed to free forage, the sites should ideally be removed from agricultural areas where exposure to pesticides may take place and confound the study.

Semi-field studies also provide an opportunity to collect exposure data. Measurement of residues in pollen and nectar as well as on/in foliage is recommended to provide data to confirm the level of potential exposure to test hives as well as to refine RQs as part of the Tier I risk characterization. Therefore, consideration should be given to recommending residue data in pollen and nectar of treated plants in semi-field studies provided that such measurements do not interfere with the exposure to bees. Depending on the crop treated, samples of pollen and/or nectar may not be readily collected from treated plants and it may be necessary to consider alternative means of collecting these samples. One option may be to use the confined bees themselves to collect samples of pollen/nectar; otherwise, it may be necessary to use intact flowers as an indirect means of determining residues in pollen/nectar combined. Although the endpoint of a semi-field study can be represented in terms of an application rate, measuring pesticide residues in pollen and nectar allows for a more direct understanding of the concentrations to which the test bees are exposed and allows for transferability of the effect results of the study to other crops, for which empirical residue data may be available for pollen and nectar. Additionally, pollen and nectar (stored in hives as bee bread and honey or uncapped honey) may be sampled within the hive to further characterize the exposure to colony. For systemic compounds, application timing represents an uncertainty as movement of the compound to pollen and nectar varies with the plant species and cultivation techniques.

As alluded to previously, semi-field studies conducted in tunnels/enclosures have limitations as the colonies can typically be maintained in the enclosures for only limited amounts of time before food reserves are exhausted and bees are stressed by the confined conditions. In addition, the foraging is not entirely natural because the tunnels may alter the foraging behavior of the bees. These stresses may impact the function of the colonies over time and result in adverse impacts, such as reduced brood production or decreased food storage, from the enclosure itself. As with all higher tier studies, the statistical power of tunnel studies is often low due to high variability as well as low replication; however, as with any study, if sufficient resources are available, the number of replicates may be increased to increase the statistical power.

Tier II tunnel studies provide relevant toxicity information, given the constraints noted above, from a limited exposure duration to characterize risk associated with specific use patterns based on the selected tunnel design. Not all crops bloom for extended periods, and some exposure periods will be short in duration as in the case of a foliar application with a relatively short foliar dissipation half-life. So the tunnel study is most appropriate for assessing scenarios in which a short exposure period (*e.g.*, short bloom period or rapid degradation/dissipation) is anticipated. Multiple application rates may be tested for a specific crop to provide a NOAEC. This application rate may then be compared to proposed application rates to characterize the potential risk from exposure to a specific application use pattern or determine the utility of various mitigation options. In addition, the measurement of residues in pollen and nectar associated with the various application rates tested in the tunnel study may provide information relevant to other crops when data are available from targeted pollen and nectar residue studies. It is important to note, however, differences between the crop tested in the tunnel and the actual uses (*e.g.*, the proposed crop uses on the label) to which the tunnel study results are intended to represent; these differences may include bee attractiveness, pollen and/or nectar production, bloom duration, *etc.*, in order to identify if a comparison is appropriate.

While tunnel studies have limited value in addressing uncertainties related to long-term chronic exposure and long-term chronic effects, they can provide valuable information on shorter exposure durations including effects from all exposure routes (*e.g.*, contact and oral for foliar applications), and on actual residue exposure levels specific to application methods, rates and test crops under consideration.

3.2.5.2. Feeding Design Studies

The methodology described by Oomen et al. 1992[20], and the extended feeding field study design proposed in the SAP White Paper, may also be considered a useful study for assessing the potential effects of pesticides on bees at the colony level.

Rather than restricting bees to tunnel enclosures with a treated crop, colonies are unrestricted and fed food sources spiked with known concentration of pesticides. The amount of pesticide can be monitored to provide an estimate of an overall amount of pesticide dose consumed in hives.

The extended feeding study design offers some advantages over the tunnel design in that the duration of exposure can be extended (e.g., weeks or months compared to 10 days in tunnels) due to the lack of confinement stress caused by the tunnels.

The exposure may also be timed with a dearth period in the available food resource to maximize the potential for exposure to treated diet. This design may be especially applicable to pesticides whose use pattern and fate properties suggest that prolonged exposure of bees may be likely.

Some limitations of the feeding design include the fact that bees are free to forage on sources other than the spiked diet, which may introduce uncertainty in the exposure assessment.

Currently available data from these studies suggests that the colony will completely consume the provided sucrose solutions, but freely foraged nectar and pollen also enter the hives. In the feeding study test bees are exposed through routes defined specifically by which spiked food sources are provided, while bees in the field are likely exposed concurrently via multiple exposure routes, including both pollen and nectar depending on crop species. It is noted that feeding studies may also include spiked pollen in addition to spiked sucrose.

Test bees are also typically exposed to a range of chosen test concentrations. Such concentrations may not represent exposure resulting from proposed application methods/use patterns. Differences in the field exposure routes and residue levels from the tested exposure routes and test concentrations may contribute to the uncertainties associated with effects observed in the study. In addition, excessive sucrose and or pollen supply may have impact on the foraging behavior of foraging bees and cause uncertainty in effects on colony development.

Similar to the tunnel studies, colonies in the field feeding studies also have high variability between colonies that decreases the statistical power of the tests; however, if sufficient resources are available, the number of replicates may be increased.

Given the above limitations, the feeding study designs may provide useful information for the characterization of risk associated with various crop use patterns. The study design can incorporate multiple treatment levels of residues in spiked food to obtain a dose response and a NOAEC at the colony level for the specific route of dietary exposure (e.g., pollen, nectar, or both) employed in the study.

The dose response from the extended feeding study may then be compared to residue levels measured in specially designed pollen and nectar residue studies that are tied to specific application rates. When evaluating these comparisons specifically for nectar, the risk assessor should consider the sugar content of the feeding solution relative to the sugar content of the nectar in a given crop as the amount of pesticide per amount of sugar may differ. In the absence of crop specific information, the Tier I exposure assumption of an average value of 30% for nectar may be assumed in order to provide a more refined comparison based on residue levels normalized by sugar content. These comparisons of the dose response with measured residue levels may provide colony-level information for the risk characterization of bee exposure to a given pesticide in a specific crop.

The extended feeding study design also offers some advantages over the tunnel design in that the duration of exposure can be extended (e.g., weeks or months compared to 10 days in tunnels) due to the lack of confinement stress caused by the tunnels.

3.2.6. Tier III Effects Characterization

As indicated in the White Paper, full-field studies represent the highest level of refinement for pollinator studies since they are intended to reflect the potential effects of a pesticide on bee colonies under actual chemical use conditions. Tier III studies may be considered when pollinator risks cannot be excluded at lower tiers of the risk assessment. These studies are intended to address specific uncertainties, *i.e.*, risk hypotheses, which have been identified through lower tier studies and/or through the open literature under reasonable worst case exposure scenarios in the field. While guidance is available for field pollinator testing (*e.g.*, OCSPP Guideline 850.3040[21], EPPO 170), this information is relatively generic and is intended to provide information on study design elements that should be considered in conducting field pollinator studies.

Additional general guidance is provided in the Field testing for Terrestrial Wildlife guidance document (OCSPP Guideline 850.2500)[22]. Because full-field studies are intended to address specific uncertainties, the study protocols should be reviewed and approved by the risk assessors prior to the conduct of the study.

Similar to the semi-field study protocols intended to address specific uncertainties, approved study protocols should be made available to other risk assessors who may be confronted with similar uncertainties in the future.

Although field pollinator testing has historically focused on honey bees, these studies are not limited to the eusocial *A. mellifera* and could include other bees (*e.g.*, bumble bees and solitary species) that are either run concurrent with or *in lieu* of honey bees. The rationale for recommending an alternative test species should be included in a proposed protocol and reviewed by the risk assessors. Given the variability associated with field-scale studies, plot size and attractiveness of the study crop should be considered. The use of extinction curves (*i.e.*, the decline in the number of foraging bees relative to distance from their colony) for estimating foraging distances of test species as a means of determining appropriate plot sizes could be useful provided such data are available.

Appendix 4. Tier 3 Field Study Design Considerations and the White Paper identify multiple study design considerations for full-field testing, but the risk hypothesis should be a primary consideration.

The duration of field studies must be weighed against the extent to which confounding effects may limit the utility of information obtained from the study. Consideration should be given to the type of pollinator effects to be addressed in the field study.

For chronic effects, a longer study duration including an overwintering component may be considered. However, as outlined in the White Paper, multiple factors (*e.g.*, nutrition, disease, pests) can impact colony survival, and longer study durations can influence the extent to which these other factors may confound the study results. Frequent manipulations of the colony to collect measurement endpoints can also stress the colony and affect the extent to which it is vulnerable to disease/pests. Therefore, the decision to collect such information should be weighed against the importance of these data in addressing the risk hypothesis.

As with semi-field studies, the full-field studies should include measures of exposure that can be used to link residues in pollen/nectar (and foliage) back to specific application rates. Measured residue values provide a means of ensuring that exposure actually occurs for durations that are expected to occur under actual use conditions of the pesticide under evaluation. These data also provide a means of tracking the movement of residues into various compartments of the intact colony and determining the extent (*i.e.*, magnitude and duration) to which other castes and/or life stages of the bees may be exposed.

In addition, measures of foraging activity on the treated crop can provide the risk assessor with information to determine the extent to which the bees utilize the treated crop.

Similar to semi-field studies, samples of pollen and/or nectar may not be readily collected from treated plants and it may be necessary to consider alternative means of collecting these samples (*e.g.*, use confined bees to collect pollen/nectar).

In addition, as bees are likely to forage on plants other than the target crop in the field, pollen composition (pollen palynology) and residue analysis of pollen collected by free-foraging bees can be used to estimate the extent to which bees are foraging on the treated and untreated crop (i.e., level of exposure dilution) in the study. Such exposure dilution and associated observed effects can allow for a more direct understanding of the actual level of exposure at which to test hives, and potential transferability of the effect observed in the study to other crops for which empirical residue data and foraging preference may be available.

As discussed in the White Paper, it is incumbent on the risk assessor to evaluate whether statistically significant measurement endpoints are also biologically significant in Tier III and in the Tier II studies discussed previously.

Similarly, high variability in some measurement endpoints may make it difficult to detect a statistically significant effect; however, a sufficient trend and magnitude of response may be reported to support the conclusion that the effect is biologically significant. The utility of these effects should be discussed in the risk characterization section of the risk assessment.

In the preceding sections, studies conducted at the screening level (Tier 1) and at high levels of refinement (Tiers II and III) are discussed. **Table 4** provides an overview of strengths and limitations of various bee toxicity studies in the context of overall bee risk assessment objectives. However, strengths and weaknesses associated with any study depend on the purpose of the study and the specific hypotheses being tested.

Table 4. Expected endpoint and consideration of the strengths and limitations of various bee toxicity studies

Study Name	Primary Endpoints	Strengths	Limitations
	Tier I		
Adult, Acute Contact	Mortality, Contact LD_{50}	• Quantifiable test doses • Dose-response curve is generated • Some sublethal effects can be measured	• Only 1 exposure route considered • NOAECs are typically not generated • Acute exposure only • Measurement of sublethal effects is often limited in scope • Effects are assessed at the individual level
Adult, Acute Oral	Mortality, Oral LD_{50}		
Adult, Chronic Oral	Mortality, NOAEC	• Quantifiable test doses • NOAEC and/or doseresponse curve can be generated • Some sublethal effects can be measured	• Only 1 exposure route considered • Measurement of sublethal effects is often limited in scope • Effects are assessed at the individual level • Test is currently under development
Larval Acute (single dose)	Mortality, Larval LD_{50}/NOAEC	• Quantifiable test concentrations • NOAEC and/or doseresponse curve can be generated • Contact and oral exposure routes are included • Larval effects are important for some MOAs (insect growth regulators)	• Actual consumed dose may vary • Assessment of effects through pupation is currently difficult • Effects are assessed at the individual level • Test is currently under development (chronic)
Larval Chronic (repeat dose)	Adult Emergence/NOAEC		
Foliar Residue	Mortality, Residual toxicity and/or RT_{25}	• Contact exposure through residues on foliage • Can assess pesticide residual toxicity	• Acute exposure only • Actual dose is not quantified • Effects are assessed at the individual level

Study Name	Primary Endpoints	Strengths	Limitations
	Tier 2		
Semi-field, Tunnel	Colony strength Brood pattern and development Foraging activity	• Multiple exposure routes (contact, oral) related to pesticide use methods	• Short-term exposure only (usually 7-10 days in tunnel)
	Worker mortality and behavior Food storage and consumption Queen health	• Minimizes influence of outside exposure to other chemicals • Some behavioral endpoints can be quantified • Standard test protocol (OECD) • Colony-level effects can be related to application rate and/or residues	• Foraging may not be natural • Stress on colonies from tunnel confinement • Replication and statistical power are often low • Usually based on a surrogate crop
Semi-field, Feeding	Colony strength Brood pattern and development Foraging activity Food storage and consumption Worker mortality and behavior Foraging activity Queen health	• Long-term exposure can be assessed • Colony-level effects can be related to concentration in diet • Greater replication can be achieved vs. tunnel or full field studies • Greater control over exposure vs. full field studies	• Oral route only • Consumed dose may differ from that encountered in the field • Protocols have not been standardized • Confounding influences of off-site foraging and exposure • Foraging may not be natural
	Tier 3		
Full Field (Experimental)	Colony strength Brood pattern and development Foraging activity Food storage and consumption Worker mortality and behavior Queen health	• Most environmentally realistic of crop/pesticide exposure conditions • Can resolve specific uncertainties raised from lower tiers	• Practical constraints may limit ability to assess "high end" exposure scenarios • Replication and statistical power often low • Confounding influences of off-site foraging and exposure • Costly

Table 4. (Continued)

Study Name	Primary Endpoints	Strengths	Limitations
Full Field (Monitoring)	Colony strength Brood pattern and development Food storage and consumption Worker mortality and behavior Queen health	• Most environmentally realistic of crop/pesticide exposure conditions • Can incorporate multiple crop exposure scenarios	• Exposure may be difficult to quantify • Causal linkages may be confounded by other stressors • Costly

4. Risk Characterization

Risk characterization is the final phase of the risk assessment process and consists of two parts, *i.e.*, risk estimation and risk description. Risk estimation involves integration of exposure and effects information to estimate the likelihood of adverse effects on the ecological receptors as a result of exposure. In Tier I, risk estimation involves the calculation of RQs. Risk description includes an interpretation of the risks in the context of uncertainty and sensitivity of risk estimates to underlying assumptions and quality of data. In addition, risk description considers how risk can be mitigated through restrictive label language and/or best management practices (BMPs). A weight-of-evidence approach should be used in the overall risk characterization. The risk characterization should also discuss data gaps and whether uncertainties could be readily addressed through additional data.

4.1. Risk Estimation

4.1.1. Calculation of Risk Quotients for Tier I Risk Assessment

Table 5 lists the exposure and effect estimates used in quantifying risk in Tier I. As discussed in the White Paper, risks are quantified for individual bees in Tier I; however, RQ values are not derived for colony-level effects. As is the case with other taxa, the RQ value is calculated by dividing the exposure estimate by the effect endpoint. It is important that the measurement endpoint units are reported or can be converted to units that are consistent with exposure estimates such that a risk quotient can be reasonably derived for Tier 1 screening RQ values.

4.1.2. Levels of Concern for Tier I Risk Assessment

Based on the process described in the White Paper, risk estimates, *i.e.*, RQ values are only calculated on individual adult and larval bees during the screening-level, Tier I, stage. RQ values are then compared to Levels of Concerns (LOCs). The LOCs for acute and chronic exposure are 0.4 and 1.0, respectively. If an RQ exceeds its LOC, then the chemical use being assessed poses a potential risk to insect pollinators. Since RQ values are calculated for individual bees, they can be considered applicable to solitary as well as social bees although they may be conservative estimates of risk for the latter.

Table 5. Summary of exposure and effect estimates used in deriving risk quotients for Tier I risk assessments

Measurement Endpoint	Exposure Route	Exposure Estimate+	Acute Effect Endpoint	Chronic Effect Endpoint+++
Foliar Applications				
Individual Survival (adults)	Contact	AREnglish* (2.7 μg a.i./bee) ARMetric* (2.4 μg a.i./bee)	Acute contact LD_{50}	None
Individual Survival (adults)	Diet	AREnglish * (110 μg a.i /g) (0.292 g/day) ARMetric * (98 μg a.i /g) (0.292 g/day)	Acute oral LD_{50}	Chronic adult oral NOAEL (effects to survival or longevity)
Brood size and success	Diet	AREnglish * (110 μg a.i /g) (0.124 g/day) ARMetric *(98 μg a.i /g) (0.124 g/day)	Larval LD_{50}	Chronic larval oral NOAEL (effects to adult emergence, survival,)
Soil Treatments				
Individual Survival (adults)	Diet	(Briggs EEC) * (0.292 g/day)	Acute oral LD_{50}	Chronic adult oral NOAEL (effects to survival or longevity)
Brood size and success	Diet	(Briggs EEC) * (0.124 g/day)	Larval LD_{50}	Chronic larval oral NOAEL (effects to adult emergence, survival,)
Seed Treatments				
Individual Survival (adults)	Diet	(1 μg a.i /g) * (0.292 g/day)	Acute oral LD_{50}	Chronic adult oral NOAEL (effects to survival or longevity)

Measurement Endpoint	Exposure Route	Exposure Estimate+	Acute Effect Endpoint	Chronic Effect Endpoint+++
Brood size and success	Diet	(1 µg a.i /g) * (0.124 g/day)	Larval LD_{50}	Chronic larval oral NOAEL (effects to adult emergence, survival,)
Tree Trunk Applications++				
Individual Survival (adults)	Diet	(µg a.i. applied to tree/g of foliage) * (0.292 g/day)	Acute oral LD_{50}	Chronic adult oral NOAEL (effects to survival or longevity)
Brood size and success	Diet	(µg a.i. applied to tree/g of foliage) * (0.124 g/day)	Larval LD_{50}	Chronic larval oral NOAEL (effects to adult emergence, survival,)

$AR_{English}$ = application rate in lbs a.i./A; AR_{Metric} = application rate in kg a.i./ha

+Based on food consumption rates for larvae (0.124 g/day) and adult (0.292 g/day) worker bees and concentration in pollen and nectar.

+Note that concentration estimates for tree applications are specific to the type and age of the crop to which the chemical is applied.

+++To calculate RQs for chronic effect, NOAEC can be used as the effect endpoint to compare with the exposure estimate in concentration.

The LOC of 0.4 identified in the White Paper for assessing potential risks to individual bees was considered by the FIFRA SAP to be highly conservative. The value is based on the median probit dose-response slope across acute contact and oral toxicity studies. While the SAP correctly noted that the 10% level of mortality resulting from RQ values equal to an LOC of 0.4 has not been demonstrated as detrimental to intact colonies, this LOC was based on an effect level that would be consistent with background (*i.e.*, control mortality) in laboratory-based studies. This LOC is intended to be conservative and serve as a reasonable screen for determining whether higher-tier testing is needed. As discussed in the White Paper and as depicted in **Figure 1** and Figure 2, the RQ values estimated using screening-level exposure models can be refined using measured residue levels prior to determining whether additional colony-level (Tier II) data are needed. The LOC of 0.4 only applies to interpreting RQ values from laboratory-based acute toxicity studies of individual bees and does not apply to the higher-tier semi-field and full-field studies on whole colonies. The LOC of 0.4 also applies to interpreting RQ values for acute larvae toxicity since limited available data of dose-response slopes for larvae prevent determining an LOC using the same approach as for adults.

Although laboratory-based chronic toxicity studies with adult and larval bees are still under development, protocols have been drafted for conducting such studies and these data may be available for review. As discussed in the White Paper, chronic RQ values are to be compared to an LOC of 1.0, *i.e.*, risk is evaluated based on whether EECs exceed the NOAEC from chronic toxicity studies with individual bees.

For those chemicals where RQ values exceed LOCs even after exposure estimates have been refined using measured residue values in pollen and nectar, more refined testing on honey bee colonies may be needed using Tier II semi-field studies and, depending on the nature of remaining uncertainties, Tier III full-field studies. Higher-tier studies with whole colonies are used to provide a more realistic characterization of potential adverse effects to colonies since the study design is intended to reflect actual exposure conditions. The risk assessor should use semi- and full-field studies to determine whether effects reported in laboratory-based studies on individual bees are apparent at the level of the whole colony and nature, magnitude and duration of these effects considering potential routes of exposure, and the biological relevancy of effects must be gauged, as well as sublethal effects that may not manifest in Tier I studies. To the extent possible, available estimates of exposure to colony bees through measured residues in pollen and nectar coming into the colony through the labeled use of the pesticide should be characterized relative to the reported effects.

4.2. Risk Description

4.2.1. Use of Other Lines of Evidence

The risk description phase of the Risk Characterization provides an opportunity to discuss additional lines of evidence and uncertainties regarding potential risks to bees beyond the RQ values calculated in the risk estimation.

All studies and relevant information are used to evaluate the likelihood and/or extent of risks to bees under various pesticide use scenarios. Risk assessments must provide a transparent description of assumptions and uncertainties surrounding the assessment and the lines of evidence considered. To the extent possible, the risk assessment should reference more detailed discussions regarding uncertainties surrounding these lines of evidence than contained in other documents such as the Overview Document, the White Paper and this guidance. It is not necessary to include all of the uncertainties noted in this document in a risk assessment; however, if the risk assessor thinks that an uncertainty is of particular concern for the chemical being assessed, a discussion of such uncertainty should be included in the risk assessment. In characterizing uncertainty, the risk assessor should consider the full weight-of-evidence. Application of a weight-ofevidence analysis is an integrative and interpretive process routinely used to evaluate ecological toxicity data[23] in a manner that takes into account all relevant scientific information. The analysis should consider whether the existing data provide relevant, robust and consistent evidence (*e.g.*, agreement within and among the outcomes of laboratory and semi-/full-field studies) that a chemical has the potential to adversely affect bees and at what level of biological organization (*i.e.*, individual bees or at the colony level) and under what conditions (*e.g.*, at what application rate, exposure duration). As part of this analysis, the risk assessor should consider the biological plausibility of the effect, the coherence, strength and consistency of the body of information as evaluated across all of the relevant information that are available.

Incident data from the EPA Ecological Incident Information System (EIIS), PMRA Pesticide Incident Reporting Program (IRP) and other aggregate databases (*e.g.*, the EPA Incident Data System) should be considered in the context of the certainty index applied to these data.

Open literature studies that have met the applicable agency standards for inclusion in ecological risk assessments should be discussed in the risk description as well. Such studies may provide useful information in considering the overall weight of evidence for the risk characterization. Careful consideration should be given to the relevance of endpoints derived from open literature studies in relation to the assessment endpoints of concern.

Additionally, the risk description section is a good place to incorporate potential mitigation options and describe how specific mitigations could impact risk conclusions. Through the engagement of risk managers, risk assessors can incorporate multiple lines of evidence and alternative exposure scenarios into the risk description.

4.2.2. Synthesis of Risks among Tiers

The risk description should provide a synthesis of the various levels of refinement for both exposure and effects that were considered and discuss the extent to which semi-field and, when available, full-field studies of whole colonies provide information consistent with laboratory-based studies with individual bees. For example, if acute mortality was observed for larval bees under laboratory conditions, did semi-field studies identify effects in the developing brood of colonies? For social bees while the risk to individual bees quantified in Tier I are important, a critical question is whether adverse effects occur at the level of the whole colony.

The Tier I toxicity testing with honey bees typically involves summer bees as opposed to winter bees. This is primarily due to the logistical constraints of collecting suitable numbers of winter bees without irreparably damaging the colony from which they were derived. Worker bees can be potentially replaced by the queen in summer whereas they cannot in winter. The assessment should note this distinction and that reduced survival in winter bees, which cannot be replaced and which must survive for a longer duration in order to ensure that the colony successfully overwinters, can be more detrimental to the colony survival. Therefore, the timing when potential exposure to bees may occur in the field should be considered during the course of risk assessment.

Compared to worker bees in hives that are many and typically replaced, there is only one queen in each hive. The loss of the single queen can result in the loss of the entire colony although colonies may attempt to supersede a queen that is failing. Effect on queens may be tested in Tier I level by modifying the laboratory toxicity study protocol that is currently designed for worker bee individuals. As part of the evaluation of whole colonies, consideration should be given to whether there are any data with which to determine whether the queen is functioning properly. If no direct measurements of toxicity to the queen bee are available, the condition of the brood may serve as an indirect measure of queen performance. Brood pattern (*i.e.*, spotty versus uniform), presence of eggs, and relatively uniform development of the larvae and pupae can be useful indicators. Another useful indicator of queen performance can be whether there are any supersedure (presence of queen cells intended to replace an old or failing queen) in the brood comb and queen replacement; colonies in which the queen is failing may attempt to replace (supersede) the queen. Additionally, an over-abundance of drone cells may be indicative of an unfertilized queen that will lead to colony loss.

Studies of the whole colony should avoid over manipulation of the colony, general observations of colony activity and health should be included as part of the assessment. Colonies confined to enclosures in a Tier II study will likely exhibit increased stress from a variety of factors linked to the limited space within the enclosure. These factors should be described even if they are not considered treatment-related as they can affect the power of the study to detect treatment effects. Additional descriptions of pre- and post-application levels of pest and diseases should be assessed as an indicator of bee health.

4.2.3. Risk Description on Sublethal Effects

In addition to the measurement endpoints that have direct linkages to assessment endpoints and that were used quantitatively in the risk estimation section of the risk characterization, the risk description should also include, if available, sublethal endpoints (*e.g.*, behavioral effects, proboscis extension reflex) that are associated with the chemical under evaluation. However, these sublethal endpoints often lack information on subsequent effects on survival, growth or reproduction. Although the FIFRA SAP recommended an increased use of sublethal endpoints, until suitable linkages have been developed between sublethal measurement endpoints to assessment endpoints, their use in risk assessment should remain qualitative; this is consistent with the process used for other taxa as described in the Overview Document as well as in the open literature guidance.

Observation of sublethal effects is required for Tier I laboratory toxicity studies. However, these studies are tested using individual bees. The ability of these studies to test social interactions between bees (such as tropholaxis (*i.e.*, the transfer of food between bees) and flight and/or foraging behavior is limited. However, such sublethal effects information should be considered as part of the weight of evidence in deciding whether to proceed with higher tier testing and in the overall risk characterization.

It is important that sublethal effects are described even when their association with apical endpoints (*e.g.*, colony survival) may not be apparent since the absence of such an effect may be the result of limitations in study design (*e.g.*, insufficient study duration). For example, radiofrequency identification (RFID technology) has been employed to mark and track bee movement and can help to quantify foraging activity; however, situations where marked declines in foraging activity may not have been reflected in adult and/or larval survival estimates could indicate that there were compensatory mechanisms (*e.g.*, sufficient food stores in the comb) that obscured the effect.

The extent to which forage bee and/or hive bee behavior is affected by pesticide treatments should be discussed in the context of whether other measurement endpoints were affected. While impaired behavior can affect proper functioning of the hive, it is frequently difficult to detect such effects during short-term studies. Similarly, longer-term studies can be impacted by a number of factors (*e.g.*, weather, pests, disease, and available forage) that can increase variability in measurement endpoints and make it difficult to detect statistically significant effects at the colony level. To the extent possible, where behavioral effects are reported either in studies submitted in response to testing requirements or in acceptable open literature studies, an effort should be made to determine whether these effects are evident in other endpoints that would most likely be impacted. For example, impaired foraging success may affect the quantity of food reserves in the comb and may be accompanied by decreased brood production by the queen due to a reduction in available food.

When reviewing single studies, all available information should be considered in evaluating the dynamics of colony development and potential interactions among many of the measured endpoints.

4.2.4. Use of Simulation Models

Different applications of colony simulation models for pesticides are currently under evaluation. Some of these potential applications include: estimating exposures and characterizing risks of bees exposed to pesticides when little data are available, and completing a sensitivity analysis of lethal and sublethal measurement endpoints to determine which endpoints are critical to the survival of a colony (this could be used to inform toxicity test study designs or regulatory LOCs for risk assessments). The White Paper discussed simulation models and, while the SAP agreed that such models may prove useful, none of the models available were deemed suitable at this time. However, the SAP recommended that components of existing models (*e.g.*, analytical subroutines) may be appropriate for use at this time. EPA and PMRA are working with researchers in government, academia and industry both domestically and internationally to evaluate and adapt models for their use in risk assessments intended for regulatory purposes.

4.2.5. Uncertainties

The primary sources of uncertainties for the risk assessment are related to estimating pesticide exposures to bees and effect of those exposures to bees. With respect to dietary exposure, the first source of uncertainty may be related to the extent to which the amount of food consumed by bees for the Tier I exposure estimate represents pesticide concentration in bee food sources. With respect to contact exposure, there is uncertainty as to the extent that residues on leaves and even soil may be available to bees for uptake. There is also uncertainty as to the extent to which bees may be exposed to pesticide residues through various sources of water, including puddle and plant exudates, and whether that water is ingested, used to dilute honey, or used to cool the colony. The second source of uncertainty is related to differences in bee biology and their foraging behaviors that may directly impact exposure routes and the extent to which bees forage on the treated crop. All these sources of uncertainty may be reflected in the study designs at different tiers.

4.2.5.1. Uncertainties: Tier I Exposure Estimate

In the Tier I assessment, dietary exposure is estimated based on upper-bound food consumption rates for honey bees. These rates are estimated based on laboratory studies conducted under controlled conditions. Uncertainties may exist in relation to bee physiology changes that occur naturally, for example difference between summer bees and winter bees. However, the upper-bound food consumption rates using in the screening-level assessment are expected to be conservative.

The Brigg's model is currently used to estimate Tier I EECs resulting from a soil application. There are five notable limitations to using the modified Briggs' model approach. The first is that this methodology is based on empirical data from only one species of plant. The second limitation is that the data set used to derive elements of the model is based on a limited number of chemicals that represent only two classes of pesticides (*i.e.*, *O*-methylcarbamoyloximes and substituted phenylureas). The third limitation is that this approach is based on data from non-ionic organic chemicals and may have limited utility for ionic chemicals that whose transport may not be predicted well using K_{ow} and K_{oc}. The fourth limitation of the Briggs' model is that it is based on passive transport of chemicals into xylem, therefore, this approach does not directly estimate pesticide concentrations in plants that are the result of phloem transport. The fifth limitation involves the use of estimated pesticide concentrations in vegetative plant matrix (i.e., shoots) as a surrogate for nectar and pollen. As additional data become available, the utility of the Brigg's model for providing screening-level estimates of exposure from soil applications will be assessed and the model will either be revised or more appropriate means of estimating exposure will be adopted.

4.2.5.2. Uncertainties: Use of Residue Data

Use of chemical-specific pollen and nectar residue data reduces the uncertainties associated with the methods discussed above; however, these data should be used with caution. Care should be taken to ensure that the available empirical data are representative of the registered/proposed uses of the chemical of interest and do not under-represent exposures to bees. Samples collected from free-foraging bees that may forage on non-targeted crops could underestimate the actual concentration in pollen and nectar from a treated crop as the bees may not have exclusively foraged on the treated crop.

4.2.5.3. Uncertainties: Agronomic Practices

One of the most important considerations within the agronomic practices is the use of managed pollinators for crop production. For some crops, growers will bring in managed bees to augment the pollination services of local bees if the crop requires pollination and wild bee populations are insufficient for adequate pollination. These commercially managed bees may include honey bees, bumble bees, blue orchard bees, alfalfa leafcutting bees, *etc*. When commercially managed bees are used to pollinate a crop, the potential for exposure and the magnitude of that exposure to the pollinating bees may be greatly increased.

Depending on the physical-chemical property of a pesticide and use methods, agronomic practices, such as irrigation, may affect the translocation of systemic pesticides and have effects on the residues in bee food sources.

The agronomic practices may also be related to the extent to which a particular registered use may be applied across the landscape of the use. Different use patterns may occupy varying spatial extents of coverage.

4.2.5.4. Uncertainties: Pollination Biology

Uncertainties on the exposure of pesticides to pollinators in the field are also associated with plant/crop pollination biology. Risk assessment usually encompasses a wide variety of crops or plants that have unique pollination characteristics. These plants may include ornamental annuals or perennials, trees or bushes covered under forestry uses, or annual or perennial crops. The pollination biology of each of these plants is important to consider when developing a description of the potential risk to bees. In the problem formulation stage of the assessment, information on the attractiveness of plants to pollinators covered by the proposed uses should be considered to determine whether exposure may occur and the scope of a risk assessment. At the risk description phase of the assessment, information on the pollinator attractiveness of the plant as well as acreages and application methods will help the risk assessor to determine the spatial and temporal aspect of risk to the bee pollinators identified in the problem formulation as well as potential mitigation solutions.

The pollination biology of plants relates to the intrinsic characteristics of the plant itself. These characteristics include the following considerations:

Bee visitation to the flowers of the ornamental, forestry tree, or crop: Not all plants produce flowers that are attractive sources of forage for honey bees, bumble bees, or solitary bees. Conversely, flowers of some plant species last only short period. The short blooming duration reduces the potential exposure of flower visitors.

Harvest period of the crop: While a crop or plant may produce flowers that are attractive to bees, some of these crops are harvested prior to bloom.

Bloom period of the crop: Different plant species will bloom at different times of the year. In addition, the length of the bloom period can differ between plants. Some plants bloom within a specific, relatively narrow window of time called a determinate bloom period. Other crops may produce blossoms continuously over the course of the growing season or for an extended period of time, which is called indeterminate bloom. Indeterminate blooming crops provide a much longer window of potential exposure to pollinating bees.

Pollen versus nectar as the food source of the bees: Not all flowers produce a nectar reward for bees, while for some species of plants, flowers may not be the only source of nectar.(*e.g.*, cotton). Depending on the species of plant, some flowers do not produce pollen that bees will collect. For example, some varieties of citrus produce only minimal quantities of pollen, while at the same time provide a rich source of nectar for pollinating bees. On the other hand, corn is a grass monocot that produces copious amounts of pollen that bees have been shown to use to varying extents, while it produces virtually no nectar. The type of food source produced by the plant may be an important consideration for the specific crop.

Bee diversity: In terms of the types of bees, honey bees and bumble bees are colonial while there are a variety of bees, both managed and wild, that are solitary and, depending on the plant, their foraging strategies may differ substantially; therefore, potential exposure may differ.

4.2.5.5. Uncertainties: Differences in Bee Life History

As noted in the White Paper and as discussed in the SAP's response, there is uncertainty regarding the extent to which any risk assessment process that relies on data on a specific species, *e.g.*, *A. mellifera*, can be considered representative of an entire taxon or multiple taxa. This is especially true for honey bees, which are a highly social (eusocial) species, where the colony/hive is dependent on the collective tasks of multiple castes and function as a "superorganism"; whereas, the majority of other bee species, particularly those species native to North America, are solitary.

4.2.5.6. Uncertainties: Differences in Pests/Pathogens/Nutrition /Management

Multiple factors can influence the strength and survival of bees whether they are solitary or social. These factors, including disease, pests (e.g., mites), nutrition, bee management practices, can confound the interpretation of studies intended to examine the relationship of the test chemical to a receptor (*i.e.*, larval or adult bee). Therefore, most studies attempt to minimize the extent to which these other factors impact the study; however, higher tier studies afford less control over these other factors, and their role may become increasingly prominent as the duration of the study is extended. Although studies attempt to minimize the confounding effects of other environmental factors, there is uncertainty regarding the extent to which the effects of a chemical may be substantially different had these other factors been in place.

4.2.5.7. Uncertainties: Uncertainty in Study Designs

Data from the toxicity of residues on foliage study are used qualitatively to characterize the length of time that residues remain toxic to bees. The results of the guideline study may result in precautionary label statements similar to those discussed in the EPA Label Review Manual[24] or in guidance documents intended to reduce the potential effects of pesticides on bees (*e.g.*, Riedl et al. 2006)[25].

Studies in which colonies are provided a source of spiked food (*i.e.*, either spiked sugar solution or spiked pollen) are referred to as "feeding studies". Feeding studies are used as a means of examining the potential effects of pesticides on whole bee colonies that are free foraging; however, there are uncertainties associated with these studies. Since bees are provided a specific diet, *e.g.*, spiked sugar solution as proposed by Oomen et al. 1992[26], there is uncertainty whether the specific diet may afford limited nutritional diversity to the test bees. Although the quantity of food consumed is recorded, consumption rates within treatments can vary widely, and there is uncertainty as to the extent to which the test material is actually consumed versus stored in the comb. The richness of other food sources in the test area may also influence the actual consumption of spiked test chemicals. Although the colonies are typically positioned away from agricultural areas, the extent to which the free foraging bees may have access to plants that may be contaminated with other pesticides is uncertain. Additionally, these studies may frequently have large numbers of colonies, belonging to different treatment groups, in close proximity where bees may potentially cross between treatments through robbing and/or scavenging activity in impaired colonies; this activity may be enhanced in areas where alternative forage may be sparse. Finally, failing colonies may be robbed of resources by bees from other treatment/controls groups resulting in exposure to different doses of the test material than planned.

Tier II tunnel studies contain challenges inherent to the design itself. In these studies, a mesh enclosure is placed over the colonies and contains within its boundaries a specified area of a test crop to serve as a food source. The enclosures can impact the performance of the colonies and ultimately limit the development of the colonies either from the stress of the tunnels placed on the colonies or from inadequate forage for the bees within the tunnels. The availability of food source and impact on the behavior of test bees limit the exposure duration of a tunnel study. Although a Tier II study should be carefully designed, by considering the test hive size, exposure duration and adequacy of feed sources, to minimize the effect of tent enclosure itself, the reviewer should be attentive to the performance of the control colonies within the tunnels to ascertain if stress placed on the colonies has a notable effect on colony performance that may impact the ability to derive meaningful results from the study data.

Evaluating Tier II or full-field Tier III studies can be challenging because the study environment is relatively uncontrolled and a number of factors can impact the study. In the case of tunnel studies these factors can include the weather, disease/parasites, extent to which bees feed on the treated crop as opposed to simply storing the food, and the potential for contaminated or inadequate forage once the colonies are placed outside of the tunnel for extended evaluation. In the case of full-field studies, confounding factors can also include the weather, disease/parasites, the extent to which bees feed on the treated crop as opposed to simply storing the food, in addition to the alternative (untreated or treated with other pesticides) forage areas and the distance that worker bees can forage even during the exposure phase of the study. When confounding effects are encountered in these studies, it is incumbent on the reviewer to determine whether any information is contained in the study, given the complexity and expense of these studies. The SAP accurately noted that the longer a study is conducted, the greater the opportunity for other factors to affect study colonies; appropriate study designs and replication can provide a means of accounting for these other factors particularly where studies are conducted over an extended spatial-temporal framework.

Direct measures of effects on specific castes (*e.g.*, queens) may be limited due to inadequate numbers of test organisms or logistical constraints to isolating particular elements of the colony for study. Therefore, potential effects or lack thereof are frequently inferred based on other colony-level measures. For example, queen performance may be indirectly assessed through evaluation of egg laying, supersedures, and the extent of drone cell production. The risk assessment should be transparent in acknowledging the uncertainties associated with making such inferences.

Due to the high cost and large variation of field conditions associated with a field study, design of an appropriate field study can be challenging. Additionally due to concerns of using toxic reference compounds in open field studies, these studies do not typically include toxic reference compounds such as those used in laboratory-based and semi-field studies. Field studies also do not typically have high numbers of replicates; therefore, the statistical detection power of a study may be low However, full-field studies are conducted under more realistic exposure conditions (actual application rates) and are expected to reflect the potential impact of a pesticide in the field. Although the risk assessment considers the full weight of evidence across all of the studies evaluated, results from Tier III studies provide the most realistic understanding of effects on colonies from the proscribed use of the pesticide.

4.2.6. Data Gaps

One of the largest challenges facing risk assessors is the extent of variability and uncertainty associated with estimating the potential likelihood and magnitude of adverse effects to bees. The decision to recommend additional studies should be based on a reasonable risk hypothesis that has considered all of the available lines of evidence for a study. As indicated in this guidance, the screening-level risk assessment can be refined using chemical-specific residue data. Residue studies to support screening-level assessments should focus on a small number of pollinator-attractive crops that may serve as a good source of pollen or nectar for foraging bees and will be used to be representative of the broader list of crop uses. Additional considerations for crop selection could include the acreage of the crop that may be potentially treated in an area, or the percent of the crop treated as well as application methods. In examining other lines of evidence, risk assessors should determine whether exposure data on similarly structured chemicals may be of utility and/or whether data required by other regulatory authorities may be accessible. While a number of uses may be proposed, it may not be practical and/or reasonable to attempt to collect residue data on all crops in order to refine risk estimates. While the data used in residue studies are typically used to refine screening-level RQ values, test data can be useful in putting colony-level data into context particularly when similar data are collected during the conduct of semi- and full-field level studies. Based on the property of chemicals of interest, proposed uses, and available lines of evidence, a tiered approach should be adopted in determining what effect data are necessary in order to conduct an appropriate risk assessment for pollinators. Potential exposure routes and exposure duration, as well as level of exposure should be considered to determine the types of studies required. For example, if multiple exposures are expected, a chronic study may be needed. Higher tier studies are required when risk concerns cannot be addressed at lower tiers. In addition and if possible, higher tier studies should be designed and conducted to address specific risk concerns identified at lower tiers or through other available information (see **Section 3.2.1** for details for data requirement). As noted in this guidance, the need for additional data and the order of data recommendations are not dictated by a regimented sequence. Rather, multiple lines of evidence should be considered in conjunction with an understanding of risk management needs. An effort should be made to determine whether data are available for similarly structured chemicals with a common mode of action; these data may be able to address some uncertainties and reduce the need for additional testing. Similarly though, data for other compounds may inform decisions to deviate from the typical sequence of studies to recommend special studies to address particular uncertainties. At each step of the process, potential risk mitigation measures and management options should be considered that may reduce the need for additional studies since the likelihood of exposure has been sufficiently mitigated.

Appendix 1. Conceptual Models

A1.1. Non-systemic, Foliar Spray Applications

For non-systemic pesticides applied via foliar spray, dominant exposure routes of foraging bees include direct deposition of spray droplets onto bees, deposition onto plant surfaces (leaf, flower, pollen, nectar, extra-floral nectaries) followed by contact and/or ingestion, and inhalation of gaseous phase chemical (for highly volatile pesticides; **Figure A1.1**).

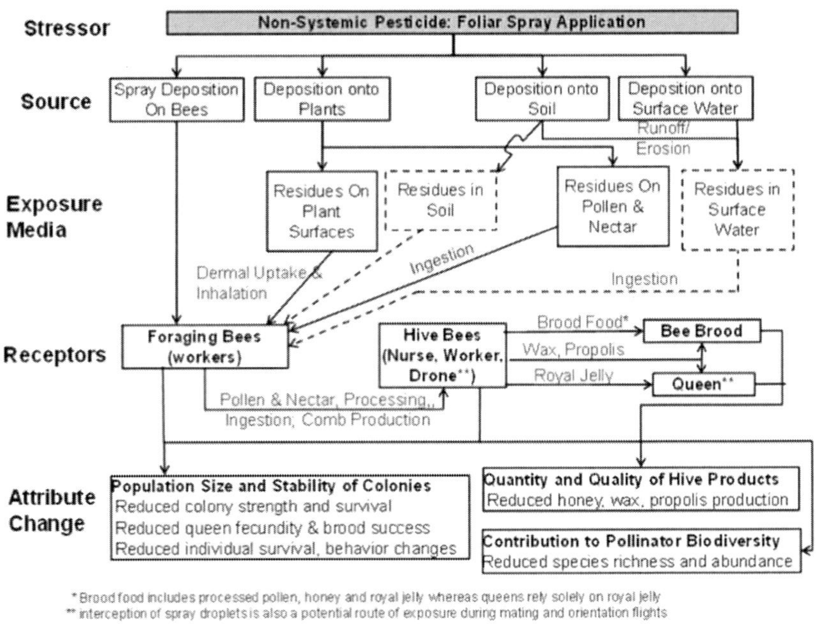

Figure A1.1. Generic Conceptual Model of Non-Systemic, Foliar-Applied Pesticides for Honey Bee Risk Assessment. Dashed lines represent routes of exposure that are not considered to be major.

A1.2. Systemic, Foliar Spray Applications

Foliar applications of systemic pesticides are likely to result in many of the same routes of exposure to honey bees as described previously for non-systemic, foliar-applied pesticides, with several important exceptions. First, deposition onto plant surfaces and soil will lead to translocation of the pesticide to other plant tissues, potentially contributing to higher quantities of pesticide residues in pollen and nectar. For persistent systemic pesticides, the exposure window could include longer periods of time (red arrows, **Figure A1.2**) compared to similar applications of non-systemic pesticides. Second, pesticide residues in plant exudates (guttation fluid, honey dew) also become a potentially relevant route of exposure.

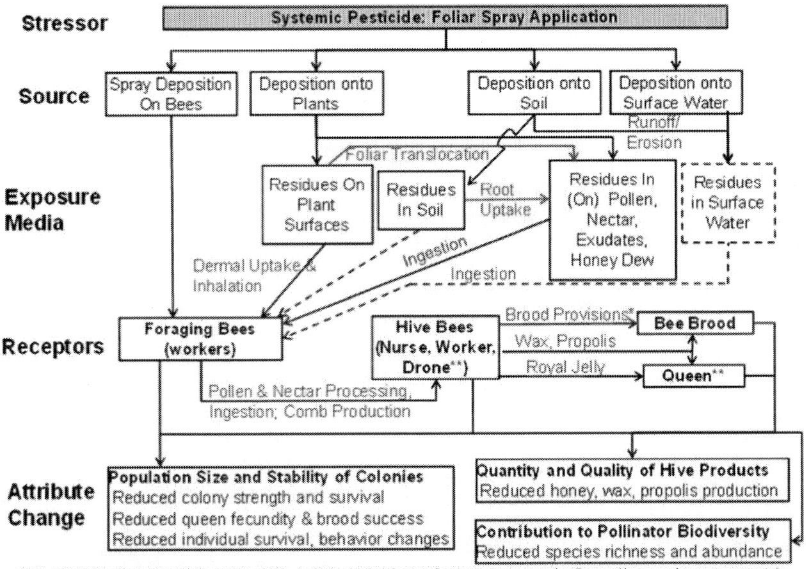

* Brood initially rely on brood jelly and royal jelly, but later in development consume processed pollen and honey; whereas queens rely solely on royal jelly.
** interception of spray droplets is also a potential route of exposure during mating and orientation flights

Figure A1.2. Generic Conceptual Model of Systemic, Foliar-Applied Pesticides for Honey Bee Risk Assessment. Red depicts systemic pathways. Dashed lines represent routes of exposure that are not considered to be major.

A1.3. Systemic, Seed Treatment

Major exposure routes of honey bees to systemic pesticides used as seed treatments include pollen, nectar, exudates (*e.g.*, guttation fluid), and honey dew resulting from translocation from the seed to growing plant tissues (**Figure A1.3**). Another important route of exposure includes contact with abraded seed coat dust during planting.

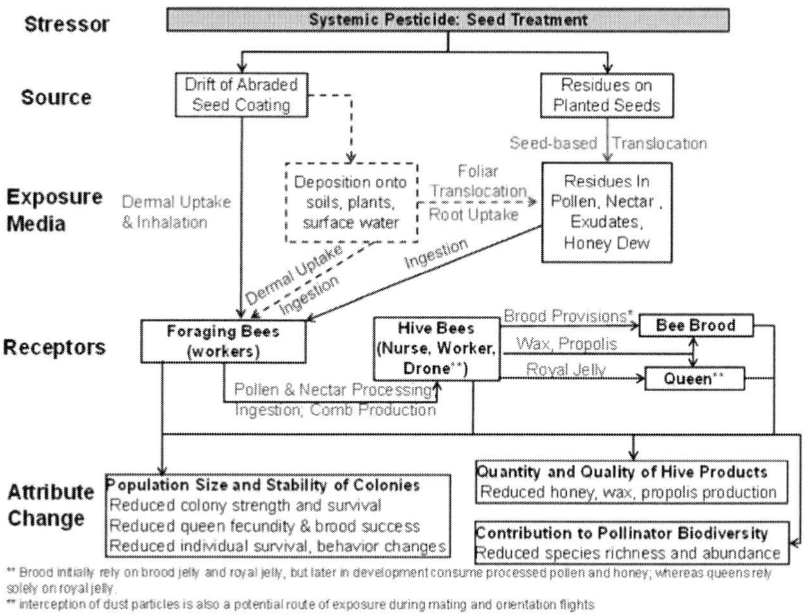

Figure A1.3. Generic Conceptual Model of Systemic, Seed Treatment-Applied Pesticides for Honey Bee Risk Assessment. Red depicts systemic pathways. Dashed lines represent routes of exposure that are not considered to be major.

A1.4. Systemic, Soil Application

Systemic pesticides are also applied soil applications (**Figure A1.4**). Exposure of honey bees to pesticides via these applications are expected to result primarily from translocation to plant tissues (pollen, nectar, exudates, and honey dew). For soil applications, there is potential exposure via runoff and subsequent translocation into plants adjacent to the treated field.

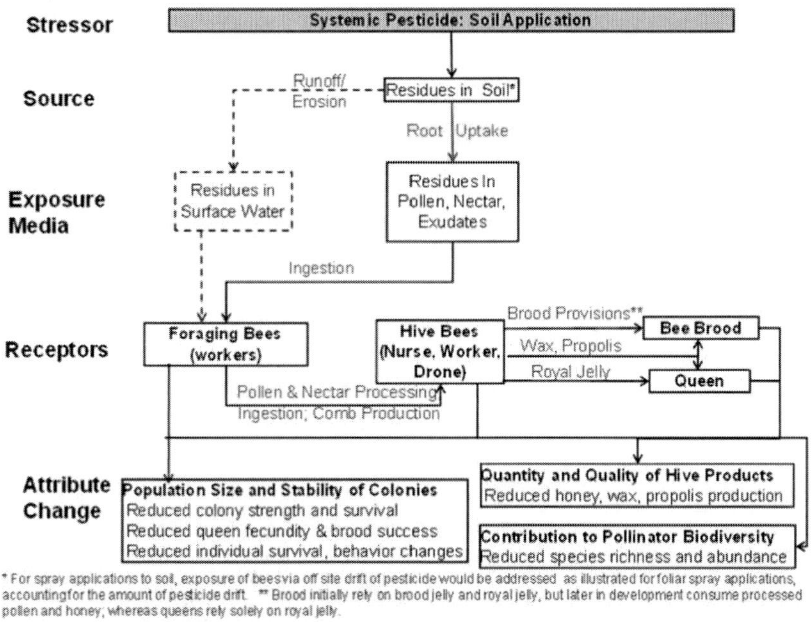

Figure A1.4. Conceptual Model of Soil-Applied Systemic Pesticides for Honey Bee Risk Assessment. Red depicts systemic pathways. Dashed lines represent routes of exposure that are not considered to be major.

APPENDIX 2. CONSIDERATIONS RELATED TO QUANTIFYING RESIDUES OF PESTICIDES IN POLLEN AND NECTAR USING PESTICIDE-SPECIFIC STUDIES

Empirical data may be used to refine conservative assumptions and reduce uncertainty associated with the Tier I exposure assessment by providing direct measurements of pesticide concentrations resulting from actual use settings. Studies investigating pesticide concentrations in pollen and nectar should be designed to provide residue data for crops and application methods of concern.

It is preferable that residues are measured from pollen and nectar samples collected directly from a crop. Otherwise they may be measured from pollen and nectar samples taken from bees that have collected the samples (*e.g.*, during Tier II tunnel studies). Measures should be taken to avoid degradation during sampling. The use of honey bees to collect samples may be desirable for crops that have flower structures that complicate the direct collection of pollen and nectar by humans. However, when honey bees are used for sampling pollen/nectar, caution should be taken to ensure that the residue data are representative of the target crop; otherwise, residue data may underestimate the target crop exposure. Chemical degradation in pollen and nectar may occur on/in bees after being collected. Bees should be limited from collecting pollen and nectar from sources other than the crop of interest.

Additional considerations may include the collection of vegetative portions of the treated crop ensuring the peak residue concentration is captured. These data, taken over a period of time, may reflect the accumulation/depuration curve of the chemical and may provide information about the transport and potential duration/concentration measures in the plant.

It is not necessary to have residue data for each crop for which a pesticide is registered/proposed. For refinement purposes, reliance on data for a select number of crops that adequately represent the diversity pollinator-attractive crops and registered uses is typically considered sufficient. In this approach, individual crops are used as surrogates for other crops. In selecting the surrogate crops, the nature of the crop (*i.e.*, whether it blooms within a specified (determinant) period or is indeterminant), its attractiveness to bees, application method, and site selection should be considered for selecting exposures that would yield a reasonably conservative estimate for exposure through pollen and nectar (e.g., higher magnitude, duration and spatial extent of residues). Regarding the number and nature of residue studies to recommend, risk assessors should consider how these data will be used in the risk assessment. To the extent possible, studies should focus on those scenarios that are believed to represent high-end exposure potential (application method/crop/soil/location). As with any study, consideration must be not only be given to the resources involved in conducting the study but also in reviewing the studies, as well as its contribution to the entire risk assessment.

The following considerations are relevant to evaluating the breadth and design of residue study protocols:

- Attractiveness of the studied crop to bees
- Spatial extent of the crop (*e.g.*, expected use acreage)
- Duration and timing of blooming period in relation to bee pollination/foraging
- Pesticide application rates, methods and timing
- Influence of soil factors and agronomic practices on residues in the crop

- Influence of soil hydric/meteorological/and transpiration conditions on crop residues
- Diversity of crop biology and physiology within a crop group, and
- The temporal nature of pesticide residues in pollen and nectar (e.g., time to peak residue occurrence and the potential for season to season accumulation of pesticide residues)

Although the aforementioned list of considerations is substantial, it most likely will not be necessary to quantify the effect of all such factors on residues in each surrogate crop. Insight into the nature of pesticide residues in crops may be obtained from magnitude of residue, plant metabolism and rotational crop studies conducted for human health assessment. Furthermore, crop residue studies can be structured so that results from earlier studies can help inform the design and need for subsequent studies.

APPENDIX 3. BEE REX

The Bee-REX model is a screening level tool that is intended for use in a Tier I risk assessment to assess exposures of bees to pesticides and to calculate risk quotients. This model is individual-based, and is not intended to assess exposures and effects at the colony-level (i.e., for honey bees).

The Tier I exposure method is intended to account for the major routes of pesticide exposure that are relevant to bees (*i.e.*, through diet and contact). Exposure routes for bees differ based on application type. In the model, bees foraging in a field treated with a pesticide through foliar spray could potentially be exposed to the pesticide through direct spray as well through consuming contaminated food. For honey bees foraging in fields treated with a pesticide through direct application to soil (e.g., drip irrigation), through seed treatments, or through tree injection, direct spray onto bees is not expected. For these application methods, pesticide exposure through consumption of residues in nectar and pollen are expected to be the dominant routes. Foraging honey bees may also be exposed to pesticides via contact with dust from seed treatments or via consumption of water from surface water, puddles, dew droplet formation on leaves and guttation fluid; however, the Bee-REX tool does not include quantification of exposures via these routes.

Appendix 3, Table 1. Summary of exposure and effect estimates used in deriving risk quotients for Tier I risk assessments

Measurement Endpoint	Exposure Route	Exposure Estimate+	Acute Effect Endpoint	Chronic Effect Endpoint+++
Foliar Applications				
Individual Survival (adults)	Contact	AREnglish* (2.7 µg a.i./bee) ARMetric* (2.4 µg a.i./bee)	Acute contact LD50	None
Individual Survival (adults)	Diet	AREnglish * (110 µg a.i /g) (0.292 g/day) ARMetric * (98 µg a.i /g) (0.292 g/day)	Acute oral LD_{50}	Chronic adult oral NOAEL (effects to survival or longevity)
Brood size and success	Diet	AREnglish * (110 µg a.i /g) (0.124 g/day) ARMetric *(98 µg a.i /g) (0.124 g/day)	Larval LD_{50}	Chronic larval oral NOAEL (effects to adult emergence, survival,)
Soil Treatments				
Individual Survival (adults)	Diet	(Briggs EEC) * (0.292 g/day)	Acute oral LD_{50}	Chronic adult oral NOAEL (effects to survival or longevity)
Brood size and success	Diet	(Briggs EEC) * (0.124 g/day)	Larval LD_{50}	Chronic larval oral NOAEL (effects to adult emergence, survival,)
Seed Treatments				
Individual Survival (adults)	Diet	(1 µg a.i /g) * (0.292 g/day)	Acute oral LD_{50}	Chronic adult oral NOAEL (effects to survival or longevity)
Brood size and success	Diet	(1 µg a.i /g) * (0.124 g/day)	Larval LD_{50}	Chronic larval oral NOAEL (effects to adult emergence, survival,)

Measurement Endpoint	Exposure Route	Exposure Estimate+	Acute Effect Endpoint	Chronic Effect Endpoint+++
Tree Trunk Applications++				
Individual Survival (adults)	Diet	(µg a.i. applied to tree/g of foliage) * (0.292 g/day)	Acute oral LD_{50}	Chronic adult oral NOAEL (effects to survival or longevity)
Brood size and success	Diet	(µg a.i. applied to tree/g of foliage) * (0.124 g/day)	Larval LD_{50}	Chronic larval oral NOAEL (effects to adult emergence, survival,)

AREnglish = application rate in lbs a.i./A; ARMetric = application rate in kg a.i./ha

+Based on food consumption rates for larvae (0.124 g/day) and adult (0.292 g/day) worker bees and concentration in pollen and nectar.

++Note that concentration estimates for tree applications are specific to the type and age of the crop to which the chemical is applied.

+++To calculate RQs for chronic effect, NOAEC can be used as the effect endpoint to compare with the exposure estimate in concentration.

Appendix 3, Table 1 summarizes the exposure and effect estimates used in developing the Tier 1 screening-level risk quotients (RQs) for individual adult bees and larvae for foliar spray applications, soil applications, seed treatments and tree trunk applications. The appropriate exposure values and effect endpoints would be used to derive these RQs. The resulting RQ values would then be compared to the LOCs for acute and chronic exposures, which are 0.4 and 1.0, respectively.

1.1. Toxicity Inputs

Estimated exposure concentrations are integrated with available toxicity data in order to characterize risks of a pesticide to honey bees. In doing so, Tier I estimated exposures and toxicity endpoints are compared based on the same exposure routes. For instance, estimated exposures through direct spray onto foraging bees are combined with toxicity endpoints from contact toxicity test, while estimated dietary exposures are matched with oral toxicity data.

For acute exposures, the endpoints are LD50 values for adult and larval bees exposed to a single dose of a pesticide and observed for several days. For adult bees, the contact LD50 value should be derived from a study conducted in a manner consistent with guideline 850.2020[27] or OECD 214. The oral exposure endpoint for adult bees should be from a study that is similar to the OECD 213[28]. For larvae, the acute oral LD50 should be from a study that is conducted in a manner consistent with OECD 237[29]. A larval chronic study is under development and endpoint for chronic toxicity should be generated in a manner consistent with the potential guidance.

At this time, standard guideline methods are not available for laboratory-based chronic tests of bees. If a scientifically-valid no-observed adverse effect level (NOAEL) or concentration (NOEAC) from a chronic study with adult bees or larvae is available, the model user can enter these into the Bee-REX tool in order to calculate chronic RQs. It should be noted that although it may be possible to extract both a NOAEL or NOEAC and an EC_x or LC_x value from the laboratory-based studies, it is important to note differences in hypothesis-based test designs needed to reliably support NOAEL/NOAEC versus regression-based study designs needed to support EC_x and LC_x estimates. When a NOAEC, or EC_x or LC_x are used as endpoints, the exposure estimates in concentrations should be used. In this case, RQs can be generated outside of the BeeREX tool by dividing the concentration based estimates of exposure by the appropriate endpoint.

It should be noted that the tool uses toxicity data for worker bees to calculate RQ values for queens and drones. There is uncertainty in using worker bees as surrogates for the royal cast due to the size differences in the bees (i.e., queens and drone adults are larger than workers).

1.2. Food Consumption Rates

As discussed in the effects characterization, oral toxicity data are necessary for adult and larvae in order to characterize the risks of a pesticide. It is important that both exposure and toxicity endpoints are in the same unit for RQ calculation; otherwise unit conversion must be conducted. When toxicity endpoints are expressed as concentration (*e.g.*, NOAEC), exposure estimates should be in the same concentration units, and conversion using food consumption rate is not needed.

When toxicity endpoints are expressed on a dose basis (i.e., μg a.i./bee), it is necessary to convert estimated concentrations of pesticides in food (expressed as mg a.i./kg) into doses. The major nutritional requirements for honey bees are met through consumption of nectar, honey, pollen and bee bread as well as royal jelly and brood food. Pesticide doses received by bees can be calculated using nectar and pollen consumption rates for larval and adult worker bees. For larvae, the proposed total food consumption rate is 124 mg/day, which is based on the total daily consumption of pollen and nectar by larvae during the last days in the life stage. For adult worker bees, the proposed food consumption rate is 292 mg/day, based on nectar consumption rates of nectar foraging bees, which are expected to receive the highest dietary exposures among different types of worker bees. In addition, it is likely that these food consumption rates are protective of drones and queens.

The Bee-REX tool calculates dietary exposure values for larvae of different ages, adult workers with different tasks (and associated energetic requirements) and royal castes. The most conservative RQ is selected for the Tier I screen. All of the RQs can be used for risk characterization purposes. Those food consumption rates, which are provided in **Appendix 3, Table 2**, are based on work described in Appendix 1 of USEPA, PMRA and CADPR 2012[30] and updated by Garber (unpublished)[31].

The tool also has the ability to override estimated concentrations with empirical data for pesticide concentrations in pollen, nectar, jelly and bee bread. When pesticide concentrations in pollen are an order of magnitude greater or more than in nectar, the most conservative adult may be the nurse bee, which consumes 1.3-12 mg pollen/day, combined with 113-167 mg nectar/day (**Appendix 3, Table 2**). The BeeREX tool uses food consumption rates that are meant to represent typical bees within a worker group or royal caste. For nurse bees, the food consumption rates of 9.6 mg pollen/day and 140 mg nectar/day were used to represent central tendencies in food consumption. The pollen consumption rate reflects the highest average daily consumption rate measured for nurse bees given the limited availability of data (only two empirical studies were located). If the refined RQ values (derived using empirical concentration data) are close to the LOC, the risk assessor may explore the influence in the variability in concentrations and in food consumption rates on risk conclusions (*e.g.*, by bounding the exposure using high and low food consumption values combined with high and low pesticide concentrations).

Appendix 3, Table 2. Estimated food consumption rates of bees

Life Stage	Caste (task in hive[a])	Average age (in days)[a]	Daily consumption rate (mg/day)			
			Jelly	Nectar[b]	Pollen	Total
Larval	Worker	1	1.9	0	0	1.9
		2	9.4	0	0	9.4
		3	19	0	0	19
		4	0	60[c]	1.8[d]	62
		5	0	120[c]	3.6[d]	124
	Drone	6+	0	130	3.6	134
	Queen	1	1.9	0	0	1.9
		2	9.4	0	0	9.4
		3	23	0	0	23
		4+	141	0	0	141
Adult	Worker (cell cleaning and capping)	0-10	0	60[f]	1.3 - 12[g,h]	61 - 72
	Worker (brood and queen tending, nurse bees)	6-17	0	113 - 167[f]	1.3 - 12[g,h]	114 - 179
	Worker (comb building, cleaning and food handling)	11-18	0	60[f]	1.7[g]	62
	Worker (foraging for pollen)	>18	0	35 - 52[f]	0.041[g]	35 - 52
	Worker (foraging for nectar)	>18	0	292 (median)[c]	0.041[g]	292

Life Stage	Caste (task in hive[a])	Average age (in days)[a]	Daily consumption rate (mg/day)			
			Jelly	Nectar[b]	Pollen	Total
	Worker (maintenance of hive in winter)	0-90	0	29f	2g	31
	Drone	>10	0	133 - 337[c]	0.0002[c]	133 - 337
	Queen (laying 1500 eggs/day)	Entire lifestage	525	0	0	525

[a]Winston (1987)

[b]Consumption of honey is converted to nectar-equivalents using sugar contents of honey and nectar.

[c] Calculated as described in this paper.

[d]Simpson (1955) and Babendreier et al. (2004)

[e] Pollen consumption rates for drone larvae are unknown. Pollen consumption rates for worker larvae are used as a surrogate.

[f]Based on sugar consumption rates of Rortais et al. (2005). Assumes that average sugar content of nectar is 30%.

[g] Crailsheim et al. (1992, 1993)

[h] Pain and Maugenet (1966)

In the proposed approach for representing food consumption rates of larvae and adults, it is assumed that exposures through consumption of nectar and pollen are conservative representations of potential exposures through consumption of honey and bee bread, respectively. This approach is likely to be conservative because it assumes that pesticides do not degrade while honey and bee bread are stored in the hive. For bees that consume honey, it is assumed that the estimated pesticide exposures can be related back to the original concentration in nectar by accounting for the amount of sugar consumed by bees. It is also assumed that pollen and nectar consumption rates and resulting exposures are protective of exposures of bees to pesticides through consumption of royal jelly and brood food. This is supported by work by Davis and Shuel 1988[32] and Kamel et al. (unpublished)[33] demonstrating that pesticide concentrations in food consumed by nurse bees are 2-4 orders of magnitude higher than concentrations measured in royal jelly. Based on these observations, for calculating RQ for queens, concentrations in royal jelly are 100X less compared to the residue in nectar or pollen, whichever is greater.

1.3. Estimating Pesticide Concentrations in Pollen and Nectar

In an ideal situation, the Tier I exposure estimates for honey bees would be based on residue values measured directly in nectar and pollen of flowers sprayed with pesticides. This cannot be achieved at this time because there is an insufficient amount of data that may be used to adequately describe the distribution of pesticide residues that occur in pollen and nectar relative to pesticide application rate. As an alternative, the proposed method relies on upper-bound pesticide residue values plant leaves, which will serve as a surrogate for pollen and nectar.

Methods for estimating dietary exposures to bees differ in the nature of the estimated concentrations in pollen and nectar. For foliar spray applications, the approach involves the use of the tall grass residue value from the T-REX model (v. 1.5) as a surrogate for pesticide concentrations in nectar and pollen. For soil treatments, the method is based on a modification to a plant-soil uptake model developed by Briggs et al. 1982 and 1983, which is designed to estimate pesticide concentrations in plant shoots; the concentrations in plant shoots are proposed as a surrogate for concentrations in pollen and nectar (following systemic transport). For seed treatments, the Tier I exposure method is based on the International Commission for Plant-Bee Relationships' (ICP-BR) 1 mg a.i./kg concentration to represent an upper-bound concentration in nectar and pollen. For tree injections and trunk drenches, the method is a simplistic approach that considers the mass of the pesticide applied to a tree and the mass of the leaves of the tree.

At the Tier I level, both acute and chronic exposure estimates are represented by the highest single day exposure value. Although a time weighted estimated exposure value may be more representative of the exposure used in a chronic toxicity test, exposure occurring over at the highest residue estimate could potentially be sufficient to elicit effects while assume no degradation occurs. Therefore, in the Tier I approach, chronic exposure is conservatively represented by the highest single day estimated exposure.

1.3.1. Foliar Sprays

For dietary exposure, the method includes the use of the upper-bound residue for tall grass used in the TREX model (i.e., 110 ug a.i./g per 1 lb a.i./A or 98 ug a.i./g per 1 kg a.i./ha) as a surrogate for pesticide concentrations in pollen and nectar of flowers that are directly sprayed with a pesticide. This exposure concentration would then be converted to a dietary dose received by adult and larval worker bees using pollen and nectar consumption rates for these two life stages, which are 0.292 and 0.124 g/day, respectively.

In order to quantify contact exposures due to direct spray, the proposed upper bound value is 2.7 μg a.i./bee per 1 lb a.i./A, or 2.4 μg a.i./bee per 1 kg a.i./ha, based on data published by Koch and Weisser (1997). As with the dietary exposure, the contact exposure value can be adjusted to account for application rate.

1.3.2. Soil Treatments

For soil treatments, it is assumed that bees will be exposed via dietary consumption of pollen and nectar that are contaminated as a result of systemic transport of pesticides from soil. For these application types, it is assumed that honey bees will not be directly exposed through contact because they are not expected to be present on the surface of the soil. The method for estimating dietary exposures to bees resulting from soil treatments is based on an empirically based model developed by Briggs et al. 1982 and 1983, with modifications (referred to here as "the Briggs' Model"). This model relates the Log K_{OW} of a chemical to its concentration in plant shoots, which can be used as a surrogate for concentrations in nectar and in pollen). In comparison to the dietary-based exposure values proposed for foliar spray applications (i.e., based on the tall grass upper-bound value), the Briggs' model generates exposure values that are usually two orders of magnitude lower. In using the Briggs' model, the approach is to use Equation 1, with the 95th percentile TSCF value that is specific to an assessed pesticide's Log K_{OW} (calculated using Equation 2). It is assumed that the resulting value is equivalent to pesticide concentrations in pollen and nectar of crops receiving soil treatments of the pesticide. The estimated concentration in pollen and nectar may be converted to dietary-based exposures for adult and larval bees using the consumption rates for pollen and nectar (i.e., 292 and 120 mg/day, respectively).

Equation 1.

$$C_{stem} = \left[10^{(0.95*LogKow-2.05)} + 0.82\right] * TSCF * \left[\frac{\rho}{\theta+\rho*Koc*foc}\right] * C_{soil}$$

Where:

C_{stem} = concentration in stems (μg a.i./g plant)

C_{soil}= concentration in soil (μg a.i./g soil)
f_{oc}= fraction of organic carbon in soil
θ = soil-water content by volume (cm^3/cm^3)
ρ = soil bulk density (g-dw/cm^3)
K_{oc} = soil organic carbon-water partitioning coefficient (cm^3/g-oc or L/kg-oc)
TSCF =Transpiration Stream Concentration Factor

Equation 2. $TSCF = — 0.0648 * (Log\ Kow)^2 + 0.241 * LogKow + 0.5822$

For the parameters in **Equation 1** that define the properties of the soil, conservative values were chosen to maximize the concentration in soil pore water and thus maximize the amount of chemical available for uptake into stems. Values were chosen to be consistent with standard scenarios used to run PRZM. For the fraction of organic carbon in soil (f_{oc}), a value of 0.01 is used. A value of 1.5 g-dw/cm^3 is selected to represent bulk density (ρ). Soil water content (θ) is set to 0.2 cm^3/cm^3. **Equation 1** can be used to calculate the concentration of a chemical in stems using the parameter values above and the application rate. If it is assumed that this application rate is homogenously distributed throughout the top 6 inches (15 cm) of the treated soil (based on the assumption that the majority of the pesticide will remain in this portion of the soil), a rate of 1 lb a.i./A is equivalent to 0.50 μg a.i./g-soil (using the bulk density), a rate of 1 kg a.i./ha is equivalent to 0.45 μg a.i./g-soil (using the bulk density).

In the Tier I approach, it is assumed that all chemicals may be systemically transported. This assumption may be limited based on Log $_{Kow}$ (Ryan et al. 1988). Therefore, it could be limited to the bounds of the Log Kow values used in the empirical data set generated by the Briggs' model (i.e., for chemicals with Log K_{ow}<5; which is the bound on the data set used by Briggs). Whether a chemical is transported systemically in plants could potentially be confirmed using empirical data submitted to EPA and PMRA (e.g., plant metabolism studies); however, it would be up to pesticide registrants to submit sufficient data to demonstrate that a pesticide is not systemic.

There are five notable limitations to using the modified Briggs' model approach. The first is that this methodology is based on empirical data from only one type of plant. The second limitation is that the data set used to derive Equation 1 is based on a limited number of chemicals that represent only two classes of pesticides. The third limitation is that this approach is based on data from non-ionic organic chemicals and may have limited utility for ionic chemicals that whose transport may not be predicted well using Kow and Koc. The fourth limitation of the Briggs' model is that it is based on passive transport of chemicals into xylem, therefore, this approach does not directly estimate pesticide concentrations in plants that are the result of phloem transport. The fifth limitation involves the use of estimated pesticide concentrations in vegetative plant matrix (i.e., shoots) as a surrogate for nectar and pollen.

At this time, the modified Briggs method represents the best available approach for estimating pesticide concentrations in pollen and nectar of crops on soil treated with pesticides. The use of this approach was supported by the FIFRA SAP. As more data become available (e.g., measured pesticide residues in pollen and nectar from crops on treated soil), EPA, PMRA and CADPR will reevaluate the modified Briggs method.

1.3.3. Seed Treatments

The Tier I exposure method for seed treatments is based on the EPPO 2010 screening value of 1 mg a.i./kg in pollen and nectar of plants to which seed treatments were made. For these application types, it is assumed that bees will only be exposed through the diet. In the Tier I approach, it is assumed that all pesticides that are applied to seeds are systemic, and therefore can be transported into pollen and nectar that may be consumed by honey bees. This approach may be used for all pesticides that are applied via seed treatment, by assuming that the upper-bound exposure value for pollen and nectar is 1 mg a.i./kg. This value can be multiplied by nectar and pollen consumption for adult and larval worker bees to determine the upper-bound doses potentially received by bees. This method can be applied to all chemicals that are applied to seeds, with no need for adjustment based on application rate or chemical properties.

1.3.4. Tree Trunk Applications

The method involves estimating the concentration in the vegetative part of the treated tree (excluding woody parts) by dividing the mass of the pesticide applied to the tree by the mass of tree vegetation represented primarily by leaves, but also by flowers. The application rate should be entered as the total mass of pesticide applied per tree.

The mass of the leaves of the tree must be estimated based on the expected characteristics of the tree to which the pesticide may be applied. It is expected that the mass of leaves and flowers of a tree will vary by species and by age of the tree, therefore a standard equation is not provided here. There are various sources available through government publications, extension offices and scientific literature that provide equations for estimating the mass of tree leaves. For example, Jenkins et al. 2003[34] provides equations for various tree species that may be used to estimate the mass of foliage for hard and softwood species. Alva et al. 2003[35] provide data on masses of citrus tree leaves.

This mass of leaves should be entered as a wet weight value. In some cases, it may be necessary to convert from dry weight to wet weight. In that case a standard assumption of 80% water content in leaves may be used (this is consistent with the T-REX approach for grass and leaves).

This approach assumes that the applied pesticide is homogenously distributed in the tree's leaves and flowers and is not present in other parts of the tree. One major uncertainty associated with this approach is the estimate of the vegetative mass of the tree, which can vary greatly based on species, age, time of the year and geography. In addition, in this approach, it is assumed that 100% of the active ingredient is taken up into the tree and moved into the leaves and flowers. It is unlikely that all of the pesticide mass applied to the tree actually enters solely to the leaves.

1.4. Using Empirical Exposure Data

If empirical data are available for a chemical, these data may be entered into the Bee-REX tool to calculate RQs. User-entered empirical exposure values will automatically over-ride estimated values. This can be accomplished by selecting "yes" to the question "Are empirical residue data available?" Once that is done, the user should enter the available values for pollen, nectar and jelly.

APPENDIX 4. TIER 3 FIELD STUDY DESIGN CONSIDERATIONS

Tier III studies conducted under full-field conditions where bees are free foraging are intended to address specific uncertainties/risks that have been identified in lower-tier studies. The design of these studies will depend on the specific questions that need to be answered; therefore, it is not possible to define a single study design or specific design elements that must be incorporated into every full-field study. Below are elements that the risk assessor should consider; however, these are not intended to be prescriptive. It is incumbent on the review team to ultimately identify the study design elements that should be considered by the pesticide registrant/applicant in developing a study protocol that is responsive to the Tier III study requirement. It should also be noted that the data package submitted by the registrant may already include a series of bee studies spanning multiple tiers; it such cases, the reviewer should ensure that the studies address the uncertainties identified by review team to determine whether additional studies may be necessary.

Chronic Field Pollinator Study Design Considerations

Application Conditions

- Maximum application rate
- Minimum reapplication interval
- Maximum number of applications
- Use of formulated end-product
- Application method
  - Foliar
  - Soil treatment
  - Seed treatment
  - Combination
- Suitable weather
  - Avoid applications when rain and/or high winds are predicted.
- Season

Test crop

- Attractive to test bees
- Long bloom period to address concerns identified at lower tiers
- Large area to ensure majority of foraging on test crop
- Follow standard [local] agriculture practices

Colonies

- Package bees/new equipment to limit incidence of disease; if older colonies are used, they should be as pest/disease free as possible.
  - Colonies should not be used if they have received any chemical treatments within last 4 weeks. Colonies suspected of having American foulbrood should not be utilized. Other disease treatments should be reported.
  - Beekeeper standard practice for maintain colony health during study
  - All treatments must be uniform across study colonies.
- Queen-right (healthy queen present); sister queens for each replicate.
- Acclimation period: 2 months minimum to establish representative age distribution in newly established hives;
- Homogeneous colony strength, brood pattern as close as possible
- If existing colonies are to be used; broad spectrum residue analysis in hive products (honey/nectar, pollen, wax); must document low incidence of diseases/parasites.
- Size of the colonies may vary depending on the focus of the study and when the study is initiated. Typically, each hive should at least 10,000 bees to cover 10 frames and include at least 5 brood frames. Excessive food storage should be avoided.
- Colonies can be positioned in plots when test crops are blooming enough to minimize test bees foraging on plants other than the test crop, *e.g.*, 20-25% bloom

Study Design Considerations

Historically there has been difficulty in controlling the extent to which the free-foraging bees utilize the treated crop or that treatment groups cross-over. Sufficiently large field plots, if feasible, will overcome the cross-over issue between plots and problems of ensuring exposure to treated crops due to competing vegetation.

- Mean honey bee foraging distance 1.5 – 3 km with extreme distances of 10 km; average surface area range 7 – 100 km^2 (Medrzycki et al.

2013[36]).
  - EFSA 2013[37] recommends minimum of 2 ha to provide sufficient flowers and support exclusive foraging; Medrzycki et al. 2013 recommends minimum of 5 ha.
- Suitable crop that is representative of actual use; good source of both pollen and nectar, (*e.g.*, phacelia (*Phacelia tanacetifolia*), oilseed rape, mustard, buckwheat)
  - Pollinator-attractive
- Account for crops/alternative forage within 3 km of colonies.

Distance of treated crop from other nectar producing plants is essential to ensure exposure and must be documented.

- Pollen traps should be used to demonstrate extent to which bees have foraged on treated crop.
- Pollen identification (palynological analysis) may be used to ensure origin of pollen
- Pollen/nectar collection for residue analyses
  - Collected by bees and sampled using pollen traps (corbicular pollen)
  - Collected directly from plants
  - Sampling of nectar forager honey stomachs
  - Sampling comb pollen/nectar

Minimum number of replicate colonies: 6 - 10 per treatment (Medrzycki et al. 2013); the number of replicates per treatment will depend on the targeted magnitude of effects and desired statistical power.

Study duration should assess at least two brood cycles (42 days) to ensure brood is exposed to residues stored in the colony (EFSA 2013).

Measurement Endpoints: depend on the risk hypothesis tested and the nature of uncertainties identified in lower-tier tests. Possible measurement endpoints may include. Adult Forage Bees

- Adult bee survival/longevity
- Adult bee foraging activity (visual counts of returning foragers; mark-and-recapture; calibrate Dead Bee Dead Zone traps)

Queen status over the course of the exposure Colony health (disease/pest incidence)

Colony Strength

- Brood (quantify number of eggs; larvae, capped cells, pollen, honey/nectar cells)
- Monitoring of brood in a minimum of two staggered cohorts, mid-way and late in the exposure period
- Adult longevity: measured by using 30 newly-emerged adult bees from each colony (minimum n=6 colonies/treatment) in a controlled laboratory cage experiment monitoring daily mortality.
- Newly-emerged bee weights
- Hive weight

Other potential endpoints

- Overwintering Success
- Fitness measure: Pathogen challenge (*e.g.*, Nosema exposure) to newly emerged bees
- Assess the ability of colonies to re-queen themselves by removing all queens and determining the success of each colony in rearing a replacement queen.

Documenting Exposure

- Residue analysis in pollen/nectar collected from crop
- Residue analyses in pollen/nectar collected from bees
- Residue analysis in pollen/nectar/honey collected from hive
- Residue analyses in bee carcasses
- Residue analyses in wax
- Foliar Residue analysis
- Measure total residues of concern (parent + degradate(s))

Suitable control bees (residue analyses to demonstrate lack of exposure). Utility of mark-and-recapture to document drift of bees from treated colonies.

End Notes

[1] USEPA. 2012. White Paper in Support of the Proposed Risk Assessment Process for Bees. Submitted to the FIFRA Scientific Advisory Panel for Review and Comment September 11 – 14, 2012. Office of Chemical Safety and Pollution Prevention Office of Pesticide Programs Environmental Fate and Effects Division, Environmental Protection Agency, Washington DC; Environmental Assessment Directorate, Pest Management Regulatory Agency, Health Canada, Ottawa, CN; California Department of Pesticide Regulation http://www.regulations.gov/#!documentDetail;D=EPA-HQ-OPP-2012-0543-0004

[2] USEPA 2012. Final FIFRA SAP Pollinator Meeting Report. SAP Minutes No. 2012-06. A Set of Scientific Issues Being Considered by the Environmental Protection Agency Regarding Pollinator Risk Assessment Framework. September 11 – 14, 2012, FIFRA Scientific Advisory Panel Meeting. http://www.regulations.gov/#!documentDetail;D=EPA-HQ-OPP-2012-0543-0047

[3] USEPA. 2004. Overview of the Ecological Risk Assessment Process in the Office of Pesticide Programs, U. S. Environmental Protection Agency. Endangered and Threatened Species Effects Determinations. Office of Prevention, Pesticides and Toxic Substances Office of Pesticide Programs, Washington DC. January 23, 2004. http://www.epa.gov/espp /consultation/ecorisk-overview.pdf

[4] Depending on physical-chemical properties, systemic pesticides can move within the vascular (xylem and/or phloem) tissues to untreated tissues of the plant or remain locally distributed through extracellular movement. Therefore, a pesticide may be xylem mobile, phloem mobile, both xylem and phloem mobile, or locally systemic.

[5] This investigation includes consideration of recommendations provided by the FIFRA SAP with regard to the model used for estimating the relative importance of drinking water as a significant exposure route. An in-depth analysis of the model used to estimated pesticide concentrations in puddles is currently being completed.

[6] USEPA, PMRA, CDPR (2012) White paper in support of the proposed risk assessment process for bees. United States Environmental Protection Agency, Office of Pesticide Programs, Washington DC. Pest Management Regulatory Agency, Health Canada, Ottawa. California Department of Pesticide Regulation, Sacramento, CA.

[7] CFR40. 2011. Part 158, subpart G, §158.630 http://ecfr.gpoaccess.gov/cgi/t/text/text-idx? c=ecfr;sid=fe712efed37d095118c7637457e011b3;rgn=div5;view=text;node=40%3A23.0.1. 1.9;idno=40;cc=ecfr#40:23.0.1.1.9.7

[8] USEPA. 1996*a*. Ecological Effects Test Guidelines OPPTS 850.3020. Honey Bee Acute Contact Toxicity. EPA 712-C-96-147. http://www.epa.gov/ocspp/pubs/frs/publications /OPPTS_Harmonized/850_Ecological_Effects_Test Guidelines/Drafts/850-3020.pdf

[9] USEPA 1996*b*. Ecological Effects Test Guidelines OPPTS 850.3030. Honey Bee Toxicity of Residues on Foliage. EPA 712-C-96-148. http://www.epa.gov/ocspp/pubs/frs/publications /OPPTS_Harmonized/850_Ecological_Effects_Test Guidelines/Drafts/850-3030.pdf

[10] USEPA 1996c. Ecological Effects Test Guidelines OPPTS 850.3040. Field Testing for Pollinators. EPA 712-C-96-150. http://www.epa.gov/ocspp/pubs/frs/publications/OPPTS Harmonized/850 Ecological Effects Test Guidelines/Drafts/850-3040.pdf

[11] *Ibid* CFR40 2011.

[12] OECD. 1998*a*. OECD Guidelines for the Testing of Chemicals. Honeybees, Acute Oral Toxicity Test. 213. http://lysander.sourceoecd.org/vl=5988235/cl=12/nw=1/rpsv/cgi-bin/fulltextew.pl?prpsv=/ij/oecdj ournals/1607310x/v1n2/s14/p1.idx

[13] OECD. 2013. OECD Guidelines for Testing Chemicals. Honey bee (*Apis mellifera*) larval toxicity test, single exposure.. http://www.oecdilibrary.org/environment/test-no-237-honey-bee-apis-mellifera-larval-toxicity-test-single-exposure_9789264203723-en

[14] OECD.1998*a*. OECD Guidelines for the Testing of Chemicals. Test Number 214, Acute Contact Toxicity Test. http://www.oecdilibrary.org/environment/test-no-214-honey bees-acute-contact-toxicity-test_9789264070189-en;jsessionid=43gvto47wnue9.delta

[15] OECD 2007. Guidance document on the honey bee (*Apis mellifera*) brood test under semi-field conditions. OECD Environment, Health and Safety Publications. Series on Testing and Assessment No. 75. ENV/JM/MONO (2007) 22.

[16] EFSA. Guidance on the risk assessment of plant protection products on bees (*Apis mellifera*, *Bombus* spp. and solitary bees. EFSA Journal 2013;11(7):3295, 266 pp. doi:10.2903/j.efsa.2013.3295. Available online: www.efsa.europa.eu/efsajournal

[17] USEPA. 2012. Ecological Effects Test Guidelines OCSPP 850.3030 Honey Bee Toxicity of Residues on Foliage. EPA 712-C018. January 2012. http://www.regulations.gov/#! documentDetail;D=EPAHQ-OPPT-2009-0154-0017

[18] Nuclei colonies have been characterized as consisting of approximately 3000 brood cells respectively 750 cm' with brood in all stages, 1 good comb with honey and pollen and approximately 6000 worker bees; the ratio of brood to food (pollen/nectar) should not exceed 4:1 (OECD 2007).

[19] OECD. 2007. Guidance document on the honey bee (*Apis mellifera* L.) brood test under semi-field conditions. Series on Testing and Assessment No. 75. ENV/JM/MONO(2007)22.

[20] Oomen, P. A. A. DeRuijter and J. Van der Steen. 1992. Method for honey bee brood feeding tests with insect growth-regulating insecticides. Bul OEPP/EPPO Bulletin 22: 613 – 616.

[21] *Ibid* USEPA.1996c.

[22] USEPA 2012. Ecological Effects Test Guidelines OCSPP 850.2500: Field Testing for Terrestrial Wildlife. Office of Chemical Safety and Pollution Prevention (7101). EPA 712-C-021. January 2012. http://www.regulations.gov/#!documentDetail;D=EPAHQ-OPPT-2009-0154-0014

[23] USEPA. 1998. Guidelines for Ecological Risk Assessment. Published on May 14, 1998, Federal Register 63(93): 26846 – 26924. http://www.epa.gov/raf/publications /pdfs/ECOTXTBX.PDF

[24] USEPA. 2012. Label Review Manual. http://www.epa.gov/oppfead1/labeling/lrm/

[25] Riedl, H. E. Johansen, L. Brewer and J. Barbour. 2006. How to Reduce Bee Poisoning from Pesticides. A Pacific Northwest Extension Publication. Oregon State University, University of Idaho, and Washington State University. http://extension.oregonstate.edu/catalog /pdf/pnw/pnw591.pdf

[26] *Ibid* Oomen et al. 1992.

[27] USEPA. 2012. Ecological Effects Test Guidelines. OCSPP 850.3020 Honey Bee Acute Contact Toxicity. EPA 712-C-019. January 2012. http://www.regulations.gov/#!document Detail;D=EPA-HQ-OPPT-2009-0154-0016

[28] OECD. 1998. OECD Guidelines for the Testing of Chemicals. Test Number 213: Honey bees, Acute Oral Toxicity Test. http://www.oecdilibrary.org/environment/test-no-213-honey bees-acute-oral-toxicity-test 9789264070165-en;jsessionid=5p2ngklfmv8p4.epsilon

[29] OECD. 2013. OECD Guidelines for the Testing of Chemicals. Test Number 237: Honey bee (Apis mellifera) larval toxicity test, single exposure. http://www.oecd-ilibrary.org/environment/test-no-237-honey-bee-apis-mellifera-larval-toxicity-test-single-exposure_9789264203723-en

[30] USEPA, PMRA, and CADPR (2012) White paper in support of the proposed risk assessment process for bees. United States Environmental Protection Agency, Office of Pesticide Programs, Washington DC. Pest Management Regulatory Agency, Health Canada, Ottawa. California Department of Pesticide Regulation, Sacramento, CA.

[31] Garber, K.V. manuscript in preparation: Estimation of food consumption rates of honey bees (*Apis mellifera*) by caste and worker task for use in assessing dietary risks of pesticides.

[32] Davis, A.R. and Shuel, R.W. (1988) Distribution of C(14)-Labeled carbofuran and dimethoate in royal jelly, queen larvae and nurse honeybees. Apicologie, 19 (1), 37-50.

[33] Kamel, A.; Dively, G.; Hawthorn, D. and Pettis, J. Unpublished research examining the fate of imidacloprid and its metabolites in honey bee hives.

[34] Jenkins, J., D. Chojinacky, L. Heath, and R. Birdsey. 2003. Comprehensive database of diameter-based biomass regressions for North American tree species. General Technical Report NE-319, Forest Service, United States Department of Agriculture.

[35] Alva, A.K.; Fares, A. and H. Dou. 2003. Managing citrus trees to optimize dry mass and nutrient partitioning. Journal of Plant Nutrition, 26 (8): 1541-1559.

[36] Medrzycki, P. H. Giffard, P. Aupinel, L. P. Belzunces, M-P. Chauzat, C. Classen, M. E. Colin, T. Dupont, V. Girolami, R. Johnson, Y. LeConte, J. Lückmann, M. Marzaro, J. Pistorius, C. Porrini, A. Schur, F. Sgolastra, N. S. Delso, J van der Steen, K. Wallner, C. Alaux, D. G. Biron, N. Blot, G. Bogo, J-L Burnet, F. Delbac, M. Diogon, H. El Alaoui, B. Provost, S. Tosi and C. Vidau. 2013. Standard methods for toxicology research in Apis mellifera. Journal of Apicultural Research 52(4): http://dx.doi.org/10.3896/IBRA.1.52.4.14

[37] EFSA..Guidance on the risk assessment of plant protection products on bees (*Apis mellifera*, *Bombus* spp. and solitary bees. EFSA Journal 2013;11(7):3295, 266 pp. doi:10.2903/j.efsa.2013.3295. Available online: www.efsa.europa.eu/efsajournal

In: Bee Health
Editor: Cristina Weaver
ISBN: 978-1-63482-671-6

Chapter 4

STATEMENT OF JEFF PETTIS, RESEARCH LEADER, USDA AGRICULTURAL RESEARCH SERVICE. HEARING "TO REVIEW CURRENT RESEARCH AND APPLICATION OF MANAGEMENT STRATEGIES TO CONTROL PESTS AND DISEASES OF POLLINATORS"*

Chairman Scott, Ranking Member Schrader and members of the subcommittee, I am Dr. Jeff Pettis, Research Leader of the Bee Research Laboratory in Beltsville, Maryland, a research laboratory dedicated to honey bee health and part of the USDA Agricultural Research Service. I am pleased to appear before you to discuss a serious threat to the honey bee and thus our food security in the United States.

Ultimately, if no long-term solutions are developed to slow bee decline, consumers will pay more for the food they buy. The foods that bees are responsible for pollinating tend to be the foods that add vital nutrients, flavor and diversity to our diet: the fruits, nuts and vegetables that maintain health. Bees pollinate more than 90 crops and are responsible for $15 billion in added crop value. Over half the nation's bees are needed to pollinate almonds alone, a $3 billion crop with increasing acreage.

---

* This is an edited, reformatted and augmented version of a statement presented April 29, 2014 before the House Committee on Agriculture, Subcommittee on Horticulture, Research, Biotechnology and Foreign Agriculture.

One of the biggest problems facing honey bees and beekeepers today is the varroa mite. The varroa mite's full name is *Varroa destructor*, and it is perhaps the most aptly named parasite ever to enter this country. *Varroa destructor* is a modern honey bee plague. It has been responsible for the deaths of massive numbers of colonies both within the United States and worldwide. This mite is native to Asia where it normally parasitizes *Apis cerana,* the eastern or Asian honey bee, an entirely different species of honey bee from *Apis mellifera,* or the western honey bee, that was brought to the New World by Europeans, and on which the U.S. now depends for crop pollination. Asian honey bees have some natural defenses against the mite and consequently are rarely seriously affected by the Varroa. European honey bees, on the other hand, have been devastatingly susceptible to varroa mite damage. The simple act of feeding by Varroa, where it pierces the skin of the bee to suck blood, can introduce bacteria and weaken the immune system of bees. Varroa mites also transmit an array of destructive viruses to honey bees, such as deformed wing virus.

When *Varroa destructor* was first found in the Unites States in 1987, beekeepers managed more than 3 million colonies for crop pollination and their winter losses were typically about 10 to 15 percent. Today, beekeepers are having trouble maintaining 2.5 million managed colonies, winter losses are averaging over 30 percent a year, and the economic sustainability of beekeeping is at the tipping point. Beekeepers have identified varroa mites as a major problem. The costs of mite controls and replacing hives that only live 1-2 years, as opposed to living 3-5 years before the arrival of Varroa, are all accumulating to the point where varroa mites are making beekeeping no longer financially viable in this country.

For commercial beekeepers, there are currently only three fast-acting treatments for varroa mites: the miticides fluvalinate, coumaphos, and amitraz. While there are also a number of folk remedies and organic treatments, none work as well as these other treatments and all involve more labor and costs to apply. However, varroa mites are adapting and becoming resistant to fluvalinate and coumaphos. Some new treatments are in the pipeline but even a new effective miticide will only provide a short-term solution because it is only a matter of time before the varroa mite will adapt to that miticide as well, continuing the destructive cycle. What beekeepers truly need are long term solutions to varroa mites.

The beekeeper's best hope is research that can build better tools to reduce the size of the varroa mite problem. Researchers at USDA's scientific agencies--the Agricultural Research Service (ARS) and the National Institute of Food and Agriculture (NIFA) are on that trail right now. In ARS, scientists are working with a total budget of approximately $11 million dollars in FY2014, with approximately $3 million targeting Varroa specifically. Additional temporary funding of $1.3 million in 2013 has been provided on bee health through the Areawide Program of ARS. These funds have helped augment the base funds and allow scientists to work closely with commercial beekeepers to try and improve colony survival.

ARS scientists are developing improved best management practices to help beekeepers deal with immediate issues of overcoming varroa mites. By applying microbiological, genomic, physiological, and toxicological approaches, we are creating new tools for beekeepers to build and maintain healthy bee populations. For long-term solutions, ARS is looking to the genetics of both the mite and the honey bee. ARS has an active breeding program designed to increase resistance mechanisms in European honey bees. For example, some bees have a propensity for nest cleaning and grooming behaviors and these have been exploited in breeding programs as control measures. ARS is also working on improving epidemiological nation-wide monitoring of pest and diseases, biochemical disruption and a host of other possibilities.

NIFA is supporting extramural research, extension, and educational programming to scientists, extension specialists and educators to address declines in pollinators. Dozens of competitive and capacity grants are focused on novel strategies to manage the varroa mite, which are expected to better protect pollinators from this devastating pest. Since 2010, NIFA has awarded competitive grants on pollinator health worth an estimated $13 million dollars, including approximately $2.6 million targeting Varroa specifically. Varroa does not act alone on bee health and thus many of these projects take a holistic approach, looking into the multiple factors affecting honey bees and other pollinators. In one NIFA funded project, University of Minnesota extension specialists are assisting honey bee queen breeders in selecting for hygienic behavior, a trait that helps bees defend against varroa mites and other diseases. In another, Cornell scientists are testing the hypothesis that giving colonies smaller hives will provide the mites fewer opportunities to reproduce and this will lower the per capita level of mite infestation of the bees.

The work at USDA is part of a government-wide response to the large and ongoing declines in pollinator populations in the U.S. and world-wide. The President's FY 2015 budget proposes over $71 million for USDA alone to focus on this issue. This includes a $25 million initiative to create an

Innovation Institute on Pollination and Pollinator Health, a competitive program that will be managed by NIFA.

As a measure of the seriousness with which the varroa issue is regarded, USDA hosted a Varroa Summit in February of this year. More than 75 representatives and researchers from beekeeping organizations, agricultural commodity groups, the crop protection industry, universities and federal agencies such as APHIS, ARS, NIFA, NRCS and EPA attended to discuss research needed to solve the problem of varroa mites. The attendees identified numerous specific short-term and long-term research priorities. Most of these concerned the need to develop the underpinnings for new approaches to controlling varroa mites: finding natural biocontrol agents, developing RNA interference as a control measure, developing areawide management practices and improving best management practices, and identifying genetic markers and breeding for bee traits that will provide varroa survivability. Attendees also recognized the need for more extensive communication between researchers and beekeepers for collection of epidemiological and economic varroa mite data and for transmitting new information from researchers on techniques for controlling varroa. One potential outcome of the Varroa Summit will be an increased level of collaboration between scientists and more public-private and Federal-university partnerships.

But even if the varroa mite problem were solved today, this would not by itself solve all of the problems facing honey bees and beekeepers. In the last 20 years, a whole host of new honey bee pathogens—viruses, bacteria, fungi, mites—have entered the United States. We know that the effects of viruses in particular are significantly exacerbated when coupled with the presence of Varroa. Colony collapse disorder, a syndrome for which scientists still do not have a cause, continues to take a toll on apiaries. Exposure to pesticides in the environment may be weakening bee colonies, possibly making them more susceptible to other stresses. A lack of diversity in nectar and pollen sources may also play a major role in stressing honey bee colonies. The loss of honey bees may also reflect a much larger issue of general pollinator declines, with honey bees acting as an indicator species. The relative contributions of different stressors for CCD is not well understood and solving this problem will take an all hands on deck approach, including research, public education, increased foraging lands and public-private partnerships to address CCD and the larger loss of pollinators.

To meet today's increasing pollination demands, we need well over 3 million managed honey bee colonies in this country. To do that, we need to make beekeeping profitable again and that starts with controlling *Varroa destructor*.

Mr. Chairman, I look forward to working with you in support of honey bees and pollinator health, a vital link in U.S. food security. Thank you again for your time. I would be pleased to answer any questions you have on varroa mites and pollinator health.

In: Bee Health
Editor: Cristina Weaver
ISBN: 978-1-63482-671-6

Chapter 5

TESTIMONY OF ARTHUR D. CUMMINGS, CEO, CAPAY FARMS. HEARING "TO REVIEW CURRENT RESEARCH AND APPLICATION OF MANAGEMENT STRATEGIES TO CONTROL PESTS AND DISEASES OF POLLINATORS"*

Good morning, Mr. Chairman and Members of the Committee. Thank you for inviting me to testify this morning on this very important subject.

My name is Dan Cummings. I am the CEO of Capay Farms in Hamilton, California, where I produce almonds and walnuts. Additionally, I am the Chairman of the Almond Board of California Bee Task Force and have served as Vice Chairman of that Board. The Almond Board of California is a Federal Marketing Order administered by the Department of Agriculture. Also, I am the Chairman Emeritus of *Project Apis m.*

My almonds are delivered to Blue Diamond Growers on whose Board I serve. Blue Diamond Growers is a nonprofit farmer-owned marketing cooperative. Blue Diamond Growers is the world's largest processor and marketer of almonds, founded in 1910 and headquartered in Sacramento, California. The company obtains its supply of almonds from its member/owners and sells them to retail chains and food processing,

* This is an edited, reformatted and augmented version of testimony presented April 29, 2014 before the House Committee on Agriculture, Subcommittee on Horticulture, Research, Biotechnology and Foreign Agriculture.

confectionery and food service companies in nearly 100 nations around the world. Almonds are grown exclusively in California and are the largest tree crop in California with a value of $6 billion. In fact, over $4 billion of almonds were exported from California to the world last year alone. Almonds are California's number one agricultural export.

Nationally almonds rank in the top three consumer food items exported from the United States. Blue Diamond Growers exports for the majority of the almond growers in the State of California. Almond production continues to expand in order to supply the world. Over 80% of the world's almond supply is produced in California. Almonds are primarily grown in central California in a 400-mile area from Red Bluff, in the north, to Bakersfield, in the south.

The Honeybee is Essential for the Global Food Supply

The honeybee is essential for the global food supply. One third of our diet comes from honeybee-pollinated plants. The continuing health of the honeybee population is a matter of concern to the global agricultural community. Managed honeybees are vital to more than 90 bee-pollinated crops in the United States.

Nearly $20 billion in farm income is dependent on honeybees, directly or indirectly. Honeybee-pollinated crops include almonds, apples, cherries, melons, pumpkins, squash and sunflowers. Honeybee-pollinated seeds are also critical to cattle and livestock that ultimately feed on alfalfa.

California almond growers depend on honeybees for their livelihood. We are very concerned about the health of honeybees. Approximately 1.6 million honeybee colonies – approximately two thirds of all the commercially kept honeybees in the United States – are needed to pollinate California's almond orchards. An almond crop depends on cross-pollination. Most almond orchards have at least two compatible varieties of almonds planted. The honeybees cross-pollinate between these varieties in order to establish the crop. Without honeybees, there would be no crop.

California almonds are the first and largest crop each spring to require honeybees for pollination. Our industry partners with beekeepers with whom we share the common goal of healthy honeybees to support the future growth of almond production and other agricultural products. California almond growers are significant contributors to *Project Apis m.,* a non-profit organization that brings together representatives from the pollination and crop production industries to support research aimed at improving the beekeeping industry. I was the Chairman of the Board of *Project Apis m.* during its first six years and remain a Board member.

Blue Diamond Growers will contribute $100,000 to *Project Apis m.* this year for research dedicated to healthier honeybees. This is in conjunction with Blue Diamond Growers' introduction of several new honey almond products.

The Almond Board of California has funded honeybee research beginning in 1976. Since 1995, it has invested $2.3 million in honeybee health research. As a grower-owned cooperative, Blue Diamond Growers is the largest single contributor to the Almond Board of California. *Project Apis m.* has invested over $2.2 million on behalf of honeybees.

The Almond Board of California created a Bee Task Force in 2005 whose members include almond growers and processors, beekeepers and researchers. The purpose of this group is to make recommendations for research and effective pollination practices. This is being done to ensure a future of strong healthy hives in sufficient numbers. I was the first grower to chair this committee and still do.

Together, the Almond Board of California is partnering with the California State Beekeepers Association and *Project Apis m.* that works with landowners and managers to grow "bee pastures" during the pollen deficient winters. This project is funded through a series of grants.

51% of almond farms are less than 50 acres and over 80% of almond farms are owned and operated by families either individually or in partnership. Almonds are the earliest blooming natural food source for honeybees after wintering on supplements. Honeybees found in California's almond orchards enjoy an abundance of natural forage. As a result, hives typically increase after almond pollination and bloom. These larger hives are then "split" into smaller units by beekeepers. It is best to ensure honeybees have a diverse, season-long chain of food sources.

The Almond Board of California is also participating in a USDA grant for advancing "Integrated Crop Pollination." This approach integrates honeybees, other managed pollinators like the Blue Orchard Bee, and "bee pasture" in addition to almonds.

Because the California almond industry recognizes the essential role honeybees play in sustaining the global food supply, it and *Project Apis m.* have invested approximately $4.5 million in honeybee research. This is more money than any other U.S. commodity has invested in honeybee research. Other industry organizations have invested additional funds in honeybee research. The focus of this research is on improving the health of hives, which includes improving honeybee nutrition, managing pests and diseases effectively, restoring honeybee genetic stock diversity, helping honeybees cope with parasites and disease, and other areas related to helping improve their health and longevity.

Our research has led to several breakthroughs in maintaining honeybee health. Experts agree that beekeeping practices in the US have changed more in the last few years than in the last 20 years. The focus has been on ensuring better honeybee nutrition and the overall improvement of hive health. Honeybees need a variety of food sources in their diet for optimum health. The Almond Board of California's support was instrumental in the development of a new nutritional supplement for honeybees that beekeepers can use in the late summer and fall when natural sources of pollen are at low levels.

Research has also resulted in establishing best practices for dealing with the Varroa mite. This is a pest that emerged in the mid-1980s that attacks beehives by weakening and shortening the life span of the honeybees on which they feed. Initial feedback from beekeepers has been that those who have adopted these newer bee management practices experience improved honeybee hive health and performance.

Our industry is also part of an alliance that created a farming guide to promote reduced-risk and environmentally responsible pest management practices based on over 5 years of field data and experience in almond orchards. The health of the honeybee is a top priority in the Best Management Practices. These guidelines are shared with all growers and include recommendations to avoid applications of insecticides during bloom; and to minimize exposure by honeybees to any spray by avoiding applications when pollen is available and honeybees are feeding.

Several promising new bee research programs funded by the almond industry are underway. Over the years, dedicated research to improve honeybee genetic stock has resulted in breeding honeybees for hygienic behavior to help control diseases like the Varroa mite, which is the most serious pest of honeybees. Current research by Dr. Walter Sheppard and Sue Cobey at Washington State University is aimed at restoring genetic diversity to commercial honeybee stock. By increasing the gene pool within breeding stock, honeybees will be better able to cope with parasites and pathogens. This project has also developed safe collection and preservation techniques for honeybee stock and genetic material.

Another project extends best practices to queen honeybee breeders to assure honeybee health and genetic diversity. Dr. Marla Spivak at the

University of Minnesota leads this research. It has increased the proportion of improved stock in commercial breeding lines. It has also implemented diagnostic and integrated pest management (IPM) programs resulting in better control of bee hive pests with fewer chemicals.

A third research project being conducted by Dr. Louisa Hooven at Oregon State University builds on past research assessing the impact of fungicides on honeybees. This work is evaluating the impact of four fungicides currently used in almonds on honeybee development.

The almond industry is the largest single contributor to increasing the health of honeybees in America. It is instrumental in the development of MegaBee, a new nutritional supplement for honeybees, which can be used in the late summer and fall when natural sources of pollen are at low ebb. It has helped establish later summer-fall feeding practices that are important for strong hives. It developed integrated pest management (IPM) approaches for Varroa mite control that use fewer chemicals in the hive and new breeding techniques for improved honeybee resistance to pests and pathogens.

In fact, beekeepers who have adopted these newer honeybee management practices experience improved honeybee hive health and performance. For example, Dr. Frank Eischen of ARS/USDA is conducting an ongoing study in Kern County, which indicates that hive build up during almond bloom resulted in an average of a 27% increase in hive strength. He notes, that at the beginning of almond bloom, the hive strength averaged 11 frames of honeybees and at the end of bloom; the hive strength averaged 14 frames of honeybees. Under average weather conditions, a standard size (referred to as "strength") hive of 8 frames of honeybees at the start of the bloom will increase in size or "strength" to 10- 12 frames at the end of the almond bloom.

This improves in warm weather conditions, like we just experienced in February 2014, where a hive of 8 to 10 frames of honeybees will increase to 15 to 16 frames of honeybees. This is an increase in size or "strength" ranging from 50% and up. Further research is currently being conducted on this year's bloom and its impact on the health of honeybees.

Meanwhile, California almond growers will continue to lead in the investment in honeybee research, including honeybee nutrition, improved honeybee genetics, the effective management of pests and diseases, and the impact of pesticides.

Thank you, Mr. Chairman for holding this important hearing on this very critical subject. I will be happy answer any questions you may have.

In: Bee Health
Editor: Cristina Weaver

ISBN: 978-1-63482-671-6

Chapter 6

TESTIMONY OF JEFF STONE, EXECUTIVE DIRECTOR, OREGON ASSOCIATION OF NURSERIES. HEARING "TO REVIEW CURRENT RESEARCH AND APPLICATION OF MANAGEMENT STRATEGIES TO CONTROL PESTS AND DISEASES OF POLLINATORS"*

Chairman Scott, Ranking Member Schrader, members of the subcommittee, I am Jeff Stone and I serve as the Executive Director of the Oregon Association of Nurseries. I welcome the opportunity this morning to provide comments for your consideration relating to pollinator health.

This morning I will discuss the merits of a discussion on pollinator health and its importance to the agricultural community as well as our environment. I will address how Oregon's nursery and greenhouse industry uses neonicotinoids. I will also talk about the potential impacts to agriculture if this chemical class is restricted or banned without proper science-based facts. Lastly, I will give a little insight on how Oregon brought together stakeholders to chart out a reasoned path on this important issue.

* This is an edited, reformatted and augmented version of testimony presented April 29, 2014 before the House Committee on Agriculture, Subcommittee on Horticulture, Research, Biotechnology and Foreign Agriculture.

OREGON NURSERY INDUSTRY BACKGROUND

The nursery and greenhouse industry is the largest agricultural sector in Oregon. Oregon represents the nation's second largest nursery state with more than $744 million in sales annually. The industry is a traded sector, much like you would see in high technology or other cluster businesses. Nearly 75-percent of the nursery stock grown in our state leaves our borders – with more than half reaching markets east of the Mississippi River. Our reach extends to international markets as well. Nursery association members represent wholesale and Christmas tree growers, retailers and greenhouse operations. Nationally the horticultural industry's production, wholesale, retail, and landscape service components have annual sales of $163 billion and sustain over 1,150,000 full and part-time jobs.

As a proud part of U.S. agriculture, we certainly understand the importance of pollinators to the agricultural industry and our natural environment. We also recognize the importance of having effective pesticides with low environmental impact. Much of the debate today will be over Neonicotinoids. This chemical class, when used properly, is vital to the success of our industry.

They are important tools in defending trees, shrubs, and plants against destructive invasive species like the Japanese Beetle, Hemlock Woolly Adelgid and Asian Longhorned Beetle and employed as part of a management strategy to control chemical-resistant whitefly species.

POLLINATOR HEALTH IS CRITICAL TO THE NURSERY AND GREENHOUSE INDUSTRY

In the summer of 2013, a misapplication of pesticides on Linden trees in Wilsonville resulted in the death of 50,000 bees due to acute toxicity, or their direct contact with the insecticide. Oregon's Department of Agriculture (ODA) conducted an investigation and instituted a temporary rule restricting the use of pesticides containing the active ingredient dinotefuran. The investigation was completed and the restriction lifted in December 2013. Effective at the start of 2014 the department has imposed label language prohibiting the use of products containing dinotefuran and imidacloprid for use on trees in the *Tilia* genus, which include linden and basswood trees – these trees are highly attractive to pollinators when in flower.

The concerns around pesticide use and potential effects on bees are very important to all pesticide users, but especially those involved in agriculture. Oregon farmers depend on bees to pollinate many of their crops. They also depend on pesticides as tools to control destructive pests. Similarly, commercial beekeepers rely on healthy crops to optimize their pollination services. This means that Oregon growers and beekeepers have a lot at stake in this conversation. Both of us want to make sure that protecting bee health, and retaining pesticides as an effective tool, are not mutually exclusive.

The association conducted extensive outreach to our members – including retailers, greenhouse operators and wholesale growers - to increase awareness of the pollinator issue. We also wanted to assess the use of neonicotinoids and understand the number of licensed pesticide applicators. Beyond the dramatic headlines, the nursery industry expressed its support of the ODA action and the industry's reservations regarding an outright ban of neonicotinoids. This chemical class, first developed in the 1990s, represents advancement over other chemical classes making them safer to both human and pollinators and are used as part of pest mitigation strategies by our greenhouse and nursery members. In some cases, neonicotinoids are approved regulatory treatments for certification and interstate movement of nursery and greenhouse crops.

While seven states have made efforts to pass anti-neonicotinoid legislation, it is critical that the federal government's efforts be science-based. Congress should listen to stakeholders from the green industry, the environment community and academia. This is what we did in Oregon and we believe it could serve as a national model to give voice to disparate views while working toward a common goal – improving pollinator health. The Environmental Protection Agency's labeling program is intended to create a unified national regulatory program that prevents patchwork lawmaking by states. One standard is critical for commerce between the states.

Science and Reason Should Go Hand in Hand

Bee health is important to all of us. Nobody wants to see adverse incidents that add to the decline of bee populations. That being said, it is easy to let emotion drive the conversation. Instead, we should let science be our guide.

Based on current science, the Environmental Protection Agency (EPA) continues to allow application of neonicotinoids with appropriate guidelines. These chemistries are among the safest available to combat many pests. We encourage Congress to direct the research community to pursue its work on this issue without bias and identify the appropriate steps to alleviate environmental and pest pressures on pollinator health.

It is important to note that neonicotinoids represent a tremendous advancement over older pesticide treatment options. When used properly, neonicotinoids effectively control problem insects, while exhibiting less impact on non-target insects (including bees). Their ability to provide residual control means fewer applications and less applicator exposure. The OAN and other nursery industry leaders fear that decisions made to restrict or prohibit use of such materials, without scientific merit, will undermine research and development into new and reduced-risk materials going forward.

These calls to ban neonicotinoids continue despite a cadre of reports that suggest their role in declining bee health is small. The USDA's 2013 report on Honey Bee Health put pesticides, in general, near the bottom of the list of factors impacting bee health. The report highlighted other issues like colony management, viruses, bacteria, poor nutrition, lack of genetic diversity, and habitat loss as more impactful. The report continued to stress that, "the single most detrimental pest of honeybees" is the parasitic Varroa mite, first discovered in the U.S. in 1987." Recent reports from the Australian Governments Pesticides and Veterinary Medicines Authority (equivalent to our EPA) supported the conclusions of the USDA report. The Australian report said that even though neonicotinoid pesticides are used there, Australia has not suffered from honey bee colony declines, like those seen in Europe and the U.S.

Since reports of significant losses to bee colonies were publicized in 2006, researchers and regulators have been looking for possible causes. A Colony Collapse Disorder (CCD) Steering Committee was formed at the national level to address the concerns over bee losses. Several individuals from the Steering Committee along with Pennsylvania State University met in October 2012 for a National Stakeholders Conference on Honey Bee Health to discuss future actions to promote health and mitigate risks to managed honey bees in the U.S. In May 2013 the U.S. Department of Agriculture (USDA) and EPA released a comprehensive scientific report on honey bee health. The report concludes that there are multiple factors that play a role in honey bee colony declines. Findings from the report include:

- There are multiple diseases associated with CCD, many of which are amplified by the Varroa mite.
- Stakeholders should adopt Best Management Practices (BMPs) to enhance bee health.

- There is need to significantly improve genetic diversity in U.S. bee populations.
- Bees require increased nutritional options (forage) to lessen susceptibility to stressors.
- There should be continued research on pesticide impacts at field-relevant exposures.
- Stakeholders need greater collaboration and information sharing among stakeholders to facilitate adoption of BMPs that are critical to improving bee health.

While the current research does not point to neonicotinoids as a primary factor in bee health decline, we know that it may be tempting to restrict use for precautionary reasons. Unfortunately this approach ignores the important role these products play in managing pests that can have devastating effects on the environment. Neonicotinoids provide unique environmental, economic and public health benefits, such as:

- Effective protection against invasive species which can harm important urban landscapes, including the Emerald Ash Borer which can devastate urban forests.
- Systemic insect control not provided by other chemical classes.
- Lower impact on many non-target organisms than older chemistries, protecting natural enemies, which allows for greater use of IPM strategies.
- Effective control of disease carrying vectors.
- Extended control, which limits the needed number of applications, and therefore limits the exposure to workers.
- Control of pests that are resistant to other chemical classes

The Environmental Protection Agency (EPA) has not followed Europe's lead by suspending or banning the use of neonicotinoid pesticides. Instead, the EPA has been active on the pollinator issue by increasing the level of funding for research into integrated pest management, which has resulted in a reduction in the use of pesticides. Several studies, including a National Academy of Sciences study on the loss of pollinators, chaired by University of Illinois entomologist May Berenbaum, indicated that there is little evidence to indicate that banning this class of chemicals would have any positive effect.

The Congress and Obama Administration Should Focus on Solving the Problem

There is legislation pending before the House Agriculture Committee (H.R. 2692, the Saving America's Pollinators Act) introduced by Oregon Congressman Earl Blumenauer (D-OR) in response to the bee incident in the summer of 2013. The bill would effectively put a national moratorium on most neonicotinoid applications until an array of studies, including multi-year "residue build-up" evaluations can be completed. The bill's proposed moratorium could be lifted only if a final determination is made that the pesticides "will not cause unreasonable adverse effects on pollinators."

On February 24, 2014 the Pollinator Protection Caucus of the U.S. House of Representatives, chaired by Congressmen Denham (R-CA) and Hastings (D-FL), held a briefing on pollinator health and invited four groups to participate. AmericanHort's regulatory and legislative affairs director, Joe Bischoff, was asked to present the horticulture industry's perspective on the issue. During the briefing, Dr. Bischoff emphasized the importance of a holistic approach to research on the issue. He stressed that, "no concerned communities, including the bees themselves, would be served if we chase a red-herring and point fingers at an easy target like pesticides, for the purpose of political expediency."

When considering regulations surrounding pesticides, we feel it is important to look at what regulations are already in place. All pesticides used in Oregon must go through registration processes mandated by the U.S. Environmental Protection Agency (EPA) and Oregon Department of Agriculture (ODA). At the federal level this happens under the Federal Insecticide, Fungicide, and Rodenticide Act (FIFRA). Initial and ongoing re-registration is subject to a substantial review process. Registered products must meet the high standard of having "no unreasonable adverse effect on health or the environment." This means that the pesticides of concern in these cases have had extensive safety testing including:

- Honeybee acute contact toxicity (all outdoor use products)
- Honey bee toxicity of residues on foliage (if high acute toxicity and exposure likely)
- Field testing for pollinators (specific conditions)

While we can understand the concerns of beekeepers, and the public at large, the issue of declining bee populations unfortunately has no simple answer. In fact, research on Colony Collapse Disorder (CCD) has highlighted a complex interaction of factors that play a role in bee health. No singular cause of the problem has been found. While pesticides are often noted as one factor, they are not considered the primary one.

The Nursery industry wishes to work with the EPA to stress the message of stewardship and compliance with label instruction. However, in the meantime we have growing concerns over the EPA taking further steps on pollinator protection through administrative action which would substantially affect turf and ornamental applications. The use of "advisory label language" is understandable when faced with unforeseen circumstances. However, we are receiving signals from a variety of stakeholders that the administration is considering extending the label changes to other products that are used for foliar sprays. It is critical that Congress and the administration understand that moving the industry toward specific application methods for systemic products (such as liquid solution or the use of dry broadcast formulations) without consideration of the efficacy and available alternatives will not solve the problem of declines in pollinator health.

Oregon Can Serve As a Model in Collaboration

While the furor over the death of bees received national notoriety, the discussion in Oregon was engaged by beekeepers, environmental groups and farm organizations. The initial bill mirrored the federal bill introduced by Congressman Earl Blumenauer and would have moved neonicotinoids to a restricted use pesticide and substantially ban the use of the product in the state. Oregon House Bill 4139 could have taken a negative approach and pitting interest group against interest group - but that did not happen. Instead, stakeholders listened to one another and determined that a science-based approach to pollinator health would lead to a better solution. Over the next two years, stakeholders will roll up their sleeves and work with our land grant university (Oregon State University), legislators, and state agencies to determine the most appropriate path forward. It is critical we work with interested parties to examine how to study this issue further and create a communication effort for the general public and industry. We all benefit when we move in a reasoned manner to evaluate trends in pollinator health, including the use of best management practices.

We must acknowledge our stewardship role in using these chemistries. When we use them, we must deploy them as part of a larger management strategy, and always remember to use them only as directed by the EPA-approved label. It is important that consequences and tradeoffs be discussed and that a decision on neonicotinoids not lead to economic harm, erosion of pollinator health, or increased human safety concerns during the application of pesticides at the nursery operation.

The nursery and greenhouse industry is working through our national association, AmericanHort, to engage with various chemical and registrant associations on the neonicotinoid issue and to conduct a survey of use at a national level. We believe an expanded look at pollinator health should be conducted and the nursery and greenhouse industry should be a reasoned voice in the discussion.

The public, environmental groups and agriculture have an opportunity to set aside short-term political points and work together on pollinator health. It is my sincere hope that Congress engages in the same spirit. Perhaps by working alongside one another, we can do what is right for pollinator health, environmental stewardship and economic prosperity of our agricultural sector.

Thank you for your time and attention.

INDEX

A

access, 28, 75, 87, 153
accounting, 154, 167
acid, 85, 92, 101
adjustment, 171
adults, 108, 110, 119, 121, 125, 142, 143, 144, 162, 163, 165, 167
advancement(s), 70, 195, 196
adverse effects, 2, 38, 54, 56, 66, 69, 83, 108, 110, 112, 118, 124, 125, 126, 141, 144, 146, 155, 198
adverse weather, 36, 37
advocacy, 35, 60, 80, 81, 84
age, 121, 122, 123, 129, 143, 163, 166, 167, 171, 172, 174
agencies, 7, 26, 27, 28, 30, 41, 43, 63, 66, 76, 126, 182, 184, 199
Agricultural Research Service (ARS), v, viii, 29, 30, 41, 67, 181, 182
agricultural sector, 14, 194, 200
agriculture, 9, 16, 33, 37, 39, 64, 83, 100, 173, 193, 194, 195, 200
Alaska, 76
alfalfa, 6, 8, 10, 20, 50, 71, 118, 129, 150, 188
Algeria, 97
almonds, 8, 9, 12, 40, 63, 93, 181, 187, 188, 189, 191
annuals, 151
apex, 95
apples, 8, 20, 56, 188
appropriations, 30, 33, 34
ARS, 30, 31, 34, 40, 105, 183, 184, 191
arthropods, 124, 127
Asia, 182
asparagus, 8
assessment, 18, 52, 67, 70, 77, 80, 82, 83, 95, 107, 108, 109, 110, 111, 112, 113, 114, 116, 117, 118, 122, 128, 134, 145, 146, 147, 149, 151, 155, 159, 160, 161
assets, 27, 96
authority(s), 54, 155
awareness, 195

B

bacteria, 23, 58, 182, 184, 196
bacterium, 23
ban, 38, 76, 78, 81, 82, 83, 97, 98, 100, 195, 196, 199
base, 183
BEE, 38, 45
beekeeping management, viii, 2, 4, 5, 48, 49, 51
beetles, 5, 40
behavioral change, 63
behaviors, 89, 149, 183
beneficiaries, 36
benefits, 12, 35, 54, 64, 70, 96, 197

bias, 196
biodiversity, 3, 49, 116, 118
bioinformatics, 33
biological control, 64
biomass, 179
birds, 5, 38, 58, 60, 88
blends, 85
blood, 23, 182
bounds, 170
brain, 89, 90, 105
brain functioning, 90
Brazil, 21, 23
breakdown, 59, 79
breathing, 23
breeding, 28, 29, 31, 36, 67, 79, 84, 183, 184, 190, 191
Bureau of Land Management, 28
businesses, 27, 65, 81, 194

C

campaigns, 65
carbon, 170
cattle, 188
CCD, vii, 2, 3, 4, 16, 18, 23, 26, 28, 30, 31, 32, 34, 36, 41, 42, 43, 44, 47, 49, 50, 67, 87, 90, 91, 184, 196, 199
Census, 13, 14, 41
central nervous system, 80
certification, 36, 195
challenges, 80, 81, 153, 155
chemical industry, 63
chemical interaction, 87
chemical properties, 110, 114, 117, 125, 171, 177
children, 54
cities, 53, 76
citizens, 19, 37
classes, 149, 170, 195, 197
cleaning, 122, 123, 166, 183
climate(s), 21, 22, 42, 91
climate change, 21, 22, 42
coatings, 63
Code of Federal Regulations, 123
cognitive impairment, 89
coherence, 145
collaboration, 29, 67, 70, 184, 197
colony collapse disorder, vii, 2, 3, 4, 33, 36, 39, 43, 44, 47, 49, 97, 102, 105, 196, 199
commerce, 54, 195
commercial, 1, 3, 4, 6, 9, 10, 12, 13, 14, 17, 20, 22, 28, 33, 34, 40, 47, 48, 49, 54, 59, 63, 66, 81, 84, 118, 182, 183, 190, 191, 195
commercial crop, 9, 63
commodity, 184, 189
communication, 29, 44, 59, 67, 75, 79, 94, 95, 97, 184, 199
communication abilities, 59
community(s), 31, 188, 193, 195, 198
competition, 21, 22, 25, 28
competitors, 23
complement, 91
complex behaviors, 88
complexity, 112, 154
compliance, 199
composition, 137
compounds, 48, 52, 64, 70, 128, 132, 154, 156
conceptual model, 110, 114, 116, 118
concordance, 125
conference, 28, 43, 67, 94
confinement, 22, 23, 131, 134, 135, 139
conflict, 88, 91
Congress, v, 1, 2, 3, 5, 32, 33, 34, 38, 48, 49, 53, 82, 83, 84, 91, 94, 95, 100, 195, 198, 199, 200
connective tissue, 23
consensus, 51, 91
conservation, 2, 5, 19, 20, 28, 30, 32, 33, 35, 38
conservation programs, 28, 30, 35
consolidation, 14
construction, 27, 101
consumers, 37, 181
consumption, 114, 117, 119, 121, 122, 123, 139, 140, 143, 149, 153, 161, 163, 165, 166, 167, 169, 171, 179
consumption rates, 119, 121, 122, 123, 143, 149, 153, 163, 165, 166, 167, 169, 179

contaminated food, 161
contamination, 22, 79
control measures, 183
cooperation, 6, 50, 61
coordination, 30, 33, 88
correlation, 81
cost, 26, 37, 66, 154
cotton, 40, 151
Court of Appeals, 99
covering, 16, 36, 37, 88
crop insurance, 36
crop producers, 9, 11, 29, 61, 67, 75, 84
crop production, 1, 8, 12, 31, 150, 189
crop residue, 161
CRP, 28, 31
cues, 93
cultivation, 71, 132
cultural practices, 64
CV, 98
cycles, 175

D

damages, 79, 82, 99
data availability, 14
data collection, 14, 30, 50
data set, 149, 170
database, 41, 69, 70, 95, 179
deaths, 48, 51, 79, 182
decision-making process, 109
DEFRA, 98, 106
degradation, 133, 159, 168
Department for Environment, Food and Rural Affairs (DEFRA), 78
Department of Agriculture, 2, 3, 47, 49, 94, 97, 179, 187, 194, 196, 198
Department of the Interior, 20, 33, 34, 38, 75, 83
deposition, 156, 157
depth, 177
destruction, 23
detection, 59, 154
dew, 157, 158, 161
diet, 2, 4, 5, 8, 19, 22, 29, 48, 49, 51, 90, 117, 121, 128, 134, 139, 153, 161, 171, 181, 188, 190
digestion, 90
direct measure, 122, 147, 159
disaster, 31, 36, 37
disaster assistance, 31, 36, 37
disclosure, 76
diseases, 16, 17, 22, 23, 29, 48, 49, 51, 59, 66, 92, 147, 174, 183, 190, 191, 196
disorder, 2, 3, 4, 33, 36, 43, 47, 49, 184
distribution, 14, 19, 80, 168, 174
diversity, 4, 19, 87, 117, 118, 152, 153, 160, 181, 184, 190
division of labor, 6, 50
Domestic Policy Council, 26
dose-response relationship, 128
dosing, 127
draft, 70, 75, 97, 128
drinking water, 117, 177
dusts, 61

E

earthworms, 58, 88
ecology, 18, 33, 64
ecosystem, 64, 76
ECPA, 98
education, 30, 63
educators, 183
egg, 6, 23, 50, 154
emergency, 36, 68, 81
Emergency Assistance, 31, 36, 45
emergency relief, 36
emotion, 195
empirical studies, 166
endangered, 42
enemies, 23
energy, 90
enforcement, 29, 67
environment(s), 6, 44, 50, 53, 54, 55, 64, 69, 79, 81, 85, 112, 154, 178, 184, 193, 194, 195, 197, 198
environmental change, 19
environmental conditions, 117

environmental effects, 55
environmental factors, 84, 152
environmental impact, 194
Environmental Protection Agency, 2, 28, 48, 49, 51, 68, 99, 107, 177, 179, 195, 197, 198
environmental stress, 2, 4, 5, 48, 51
enzyme, 87
equipment, 31, 53, 55, 79, 96, 117, 174
erosion, 200
Europe, 3, 6, 20, 21, 23, 40, 49, 52, 53, 78, 88, 92, 97, 100, 196, 197
European Commission, 77, 78, 82, 83, 97, 100
European Union (EU), 71, 77, 83, 97
evidence, 4, 18, 21, 58, 81, 83, 84, 88, 91, 108, 109, 111, 112, 114, 125, 126, 141, 145, 146, 148, 154, 155, 197
Executive Order, 43
exploitation, 42
exports, 188
extinction, 136

F

false negative, 111
families, 7, 8, 50, 189
far right, 120
Farm Bill, 32, 37, 39, 44
farmers, 6, 31, 35, 61, 75, 195
farms, 13, 14, 36, 189
FDA, 37
fear, 196
federal agency, 33, 34, 49
federal facilities, 27
federal government, 27, 195
Federal Register, 45, 96, 99, 100, 101, 178
fertility, 59
field crops, 8, 58, 96
field tests, 125
financial, 31, 35
fish, 36, 58
Fish and Wildlife Service, 20, 28, 64, 75
flavor, 181
flight, 88, 147
flowers, 21, 41, 48, 52, 56, 59, 63, 117, 120, 132, 151, 168, 171, 172, 175
fluctuations, 15, 22
fluid, 86, 157, 158, 161
Food and Drug Administration, 37
food chain, 53, 55
food production, 1, 3, 47, 54
food products, 55
food safety, 80
food security, 181, 185
Food, Conservation, and Energy Act of 2008, 32, 49
forage crops, 8
force, 26, 27, 28, 33, 66, 103
formation, 161
France, 58, 92, 97, 103
fructose, 86
fruits, 8, 181
funding, 2, 3, 30, 31, 32, 34, 35, 36, 44, 48, 49, 100, 183, 197
funds, 44, 183, 189
fungi, 58, 184
fungus, 23, 105

G

GAO, 69, 95
gene pool, 190
General Services Administration (GSA), 26, 27
genetic diversity, 19, 22, 67, 190, 196, 197
genetic marker, 184
genetics, 2, 4, 5, 22, 25, 28, 48, 49, 51, 67, 183, 191
genomics, 33
genus, 20, 194
geography, 172
Georgia, 14, 16, 31
Germany, 97
grants, 31, 33, 44, 64, 100, 183, 189
granules, 77, 83
graphite, 117
grass, 120, 152, 168, 169, 172
greenhouse(s), 6, 20, 50, 77, 83, 110, 125, 193, 194, 195, 200

growth, 90, 138, 147, 178, 188
growth rate, 90
guidance, vii, 27, 33, 34, 43, 61, 63, 64, 70, 75, 95, 107, 108, 109, 116, 120, 125, 126, 128, 131, 135, 145, 147, 152, 155, 164
guidelines, 20, 30, 64, 75, 124, 127, 128, 190, 195
guttation, 86, 157, 158, 161

H

habitat(s), 2, 4, 5, 19, 22, 25, 27, 28, 31, 32, 33, 35, 38, 45, 48, 49, 51, 59, 64, 66, 70, 76, 82, 84, 196
half-life, 133
Hawaii, 20, 75, 76
hazards, 61
health effects, 84
herbicide, 85
high winds, 173
highways, 45
hiring, 34
historical data, 14
hives, 6, 15, 23, 31, 36, 37, 39, 40, 42, 53, 54, 58, 59, 61, 62, 88, 89, 90, 92, 110, 131, 132, 134, 137, 146, 174, 179, 182, 183, 189, 190, 191
hobby, 15
homeowners, 64, 65, 74, 88
homes, 22
honey bee colonies, vii, 1, 3, 4, 13, 14, 15, 16, 17, 29, 39, 41, 47, 48, 52, 59, 67, 84, 87, 118, 144, 184
host, 23, 183, 184
House, 32, 33, 40, 43, 44, 76, 82, 83, 94, 100, 101, 181, 187, 193, 198, 199
House of Representatives, 40, 83, 100, 198
housing, 84
Housing and Urban Development, 26, 66
human, 1, 8, 54, 55, 64, 69, 85, 161, 195, 200
human exposure, 54
human health, 54, 55, 64, 69, 85, 161
humidity, 89, 117
Hunter, 56
husbandry, 84, 108
hybrid, 8
hypothesis, 48, 58, 116, 118, 124, 128, 136, 155, 164, 175, 183
hypothesis test, 175

I

icon, 71
ID, 11
ideal, 13, 168
identification, 27, 66, 148, 175
identity, 37
images, 94
immune function, 59, 89
immune system, 87, 182
immunity, 90
improvements, 27
incidence, 26, 174, 176
income, 188
indirect effect, 68
indirect measure, 118, 147
individuals, 27, 146, 196
industry(s), 11, 28, 33, 34, 37, 39, 60, 66, 72, 79, 81, 84, 105, 149, 184, 188, 189, 190, 191, 193, 194, 195, 196, 198, 199, 200
infection, 89
inferences, 154
infestations, 25
information sharing, 197
infrastructure, 33, 34
ingestion, 71, 89, 131, 156
ingredients, 54, 55, 56, 58, 64, 123
initiation, 127, 132
injections, 119, 120, 121, 168
insecticide, 2, 38, 56, 61, 63, 64, 76, 78, 83, 86, 89, 92, 194
insects, 1, 3, 5, 7, 8, 9, 20, 23, 33, 38, 53, 55, 58, 59, 66, 68, 88, 93, 117, 118, 196
integration, 141
interference, 184
invertebrates, 23, 88
investment, 191

Iowa, 39
irrigation, 56, 150, 161
isolation, 109
Israel, 97
issues, 2, 4, 5, 14, 22, 31, 34, 48, 51, 72, 74, 80, 82, 99, 183, 196
Italy, 97

K

kill, 29, 43, 53, 55, 58, 59, 67, 85, 86, 92

L

labeling, 54, 63, 70, 72, 74, 75, 84, 112, 178, 195
laboratory studies, 91, 109, 149
lakes, 55
landscape(s), 9, 19, 40, 27, 78, 131, 150, 194, 197
larvae, 4, 23, 25, 39, 89, 90, 110, 119, 121, 122, 123, 124, 127, 128, 143, 144, 147, 163, 164, 165, 167, 176, 179
laws, 2, 38, 48, 76, 84
lead, 53, 63, 147, 157, 191, 197, 199, 200
learning, 59, 89
learning behavior, 59
legislation, 38, 53, 76, 82, 195, 198
legislative proposals, 32, 38
legume, 8
life cycle, 21, 108
livestock, 36, 188
loans, 37
lobbying, 81
locomotor, 90
longevity, 29, 117, 142, 143, 162, 163, 175, 176, 190
long-term memory, 59
Louisiana, 32, 64
low risk, 108
lubricants, 63, 79

M

magnitude, 111, 119, 137, 144, 150, 155, 160, 161, 165, 167, 169, 175
majority, 12, 152, 170, 173, 188
mammals, 5, 58
management, 2, 4, 5, 16, 22, 26, 27, 29, 30, 33, 34, 48, 49, 51, 53, 61, 63, 64, 66, 67, 75, 85, 93, 100, 114, 116, 141, 152, 156, 183, 184, 190, 191, 194, 196, 197, 199, 200
manipulation, 64, 147
marketing, 37, 187
Maryland, 76, 93, 94, 103, 105, 181
mass, 76, 168, 171, 172, 179
materials, 23, 62, 64, 70, 95, 196
matrix, 150, 171
matter, 182, 188
measurement(s), 87, 116, 118, 127, 128, 129, 132, 130, 133, 136, 137, 141, 147, 148, 175
media, 4, 22, 23, 39, 51
median, 110, 123, 124, 127, 128, 144, 166
Mediterranean, 120, 131
melon, 10, 40
memory, 88, 89
memory formation, 89
meta-analysis, 88
metabolism, 90, 120, 161, 170
metabolites, 86, 179
metabolized, 86
methodology, 133, 149, 170
Mexico, 7, 42
mice, 23
micrograms, 124
migration, 42
mildew, 53
Mississippi River, 194
Missouri, 100
misuse, 58
mixing, 62
models, 109, 116, 144, 148
modifications, 169
mold, 53
Montana, 14, 16

moratorium, 198
mortality, 3, 17, 37, 45, 78, 79, 81, 85, 89, 91, 124, 127, 129, 131, 139, 140, 144, 146, 176
mortality rate, 45
motor activity, 59
multiple factors, 52, 114, 125, 136, 183, 196

N

NAS, 2
National Academy of Sciences, 39, 42, 43, 51, 102, 104, 197
National Forest System, 35
National Park Service, 28
National Research Council, 2, 14, 18, 39, 43, 51, 87
National Security Council, 26
Native Americans, 6, 21
native species, 7, 19, 21, 50
natural disaster(s), 36
natural enemies, 197
natural food, 189
Natural Resources Conservation Service (NRCS), 29, 31, 35, 64, 67
nervous system, 90
Netherlands, 19
nicotine, 56
nonionic surfactants, 92
North America, 2, 4, 6, 7, 18, 19, 20, 23, 38, 39, 40, 41, 42, 43, 45, 50, 51, 80, 93, 94, 100, 101, 103, 127, 152, 179
NRC, 4, 14, 18, 23, 25, 39, 40, 42, 51, 87, 102
NRCS, 45, 184
nuclei, 130
nuisance, 53
nutrient(s), 179, 181
nutrition, 2, 4, 5, 22, 28, 35, 37, 48, 49, 51, 58, 67, 82, 90, 101, 112, 136, 152, 190, 191, 196

O

Obama, 26, 66, 94, 99, 198
Obama Administration, 26, 66, 198
OECD, 124, 125, 127, 128, 131, 139, 164, 177, 178
Office of Management and Budget, 26
officials, 78
OH, 12, 41
oil, 92
oilseed, 175
operating costs, 26
operations, 9, 12, 13, 14, 17, 39, 79, 99, 194
opportunities, 183
organic chemicals, 150, 171
organism, 64
Organization for Economic Cooperation and Development, 124
organs, 23
ornamental plants, 53, 64, 71
outreach, 30, 63, 67, 195
overwinter colony losses, vii, 2, 3, 13, 18, 47, 77

P

Pacific, 10, 13, 40, 75, 97, 178
paralysis, 23
parasite(s), 1, 3, 4, 20, 22, 23, 25, 29, 33, 43, 47, 49, 67, 84, 154, 174, 182, 190
participants, 17
pasture(s), 53, 189
pathogens, 3, 4, 20, 22, 23, 25, 28, 30, 31, 33, 42, 47, 49, 67, 184, 190, 191
pathways, 157, 158, 159
PCA, 62
percentile, 169
permit, 27, 35, 128
personal communication, 92
PES, 61, 93
pests, 1, 2, 3, 4, 5, 23, 25, 28, 30, 38, 43, 47, 48, 49, 51, 53, 56, 58, 59, 64, 66, 67, 84, 87, 92, 101, 112, 136, 148, 152, 190, 191, 195, 197

phloem, 150, 171, 177
physical activity, 90
physiology, 149, 161
pipeline, 27, 182
plants, 8, 19, 21, 22, 29, 38, 42, 53, 56, 58, 61, 64, 65, 68, 74, 77, 82, 83, 86, 101, 108, 110, 117, 118, 120, 125, 131, 132, 137, 150, 151, 153, 158, 170, 171, 174, 175, 188, 194
plausibility, 145
poison, 53
policy, 35, 53, 70, 80, 81
policy options, 53
pollination, 1, 3, 6, 8, 9, 10, 12, 19, 20, 21, 22, 31, 35, 36, 40, 48, 51, 62, 63, 73, 79, 84, 108, 110, 116, 117, 118, 125, 150, 151, 160, 182, 184, 188, 189, 195
pollinator crisis, vii, 1, 3, 47, 49
pollinator species, vii, viii, 1, 2, 3, 4, 18, 19, 21, 48, 49
population, 5, 13, 14, 18, 20, 21, 25, 34, 38, 41, 43, 49, 51, 60, 79, 82, 83, 116, 117, 118, 188
population size, 21, 117, 118
potential benefits, 71
predators, 127
preparation, 179
preservation, 190
President, 81, 94, 99, 183
President Obama, 81
prevention, 30, 64
principles, 63
producers, 11, 15, 36
professionals, 77
profit, 188
programming, 183
progress reports, 30
project, 41, 183, 189, 190, 191
propagation, 117, 118
prophylactic, 63
prosperity, 200
protection, 20, 27, 33, 35, 40, 56, 69, 75, 76, 93, 95, 96, 97, 98, 100, 116, 117, 118, 178, 179, 184, 197, 199
public education, 27, 184
public health, 74, 197
public-private partnerships, 27, 184

Q

quantification, 130, 161
questioning, 78

R

rape, 86, 175
rape seed, 86
RE, 98
realism, 108
receptors, 141
recommendations, 27, 34, 61, 64, 67, 83, 107, 125, 155, 177, 189, 190
Reform, 103
Registry, 75
regression, 128, 164
regulations, 71, 79, 177, 178, 198
relevance, 146
reliability, 22
replication, 90, 128, 130, 133, 139, 154
reproduction, 59, 79, 116, 117, 147
reputation, 82
requirements, 32, 35, 55, 63, 64, 69, 70, 71, 72, 73, 75, 76, 122, 123, 126, 127, 148, 165
researchers, 9, 18, 20, 23, 79, 86, 88, 105, 149, 184, 189, 196
reserves, 107, 132, 148
residues, 68, 71, 82, 89, 110, 111, 114, 119, 120, 122, 123, 127, 129, 131, 132, 133, 135, 136, 138, 139, 144, 149, 150, 152, 157, 159, 160, 161, 168, 171, 175, 176, 198
resistance, 2, 22, 28, 29, 38, 42, 84, 183, 191
resources, 19, 27, 30, 61, 64, 93, 109, 133, 134, 153, 160
response, 13, 15, 20, 34, 38, 55, 63, 74, 77, 82, 88, 92, 99, 103, 107, 128, 130, 135, 137, 138, 144, 148, 152, 183, 198

responsiveness, 89
restoration, 27
restrictions, 28, 53, 54, 66, 71, 76, 77, 81, 83, 91, 112
retail, 187, 194
rights, 2, 27, 38, 59
risk management, 70, 97, 107, 108, 109, 126, 155
RNA, 184
roots, 56
routes, 71, 85, 110, 114, 117, 119, 121, 131, 133, 134, 138, 139, 144, 149, 155, 156, 157, 158, 159, 161, 164
Royal Society, 40
runoff, 158

S

safety, 61, 63, 97, 198, 200
SAP, 70, 95, 107, 117, 122, 126, 133, 144, 147, 149, 152, 154, 171, 177
scarcity, 22
science, 16, 19, 42, 70, 79, 88, 100, 108, 193, 195, 199
scientific knowledge, 67
scope, 38, 84, 138, 151
seed, 8, 10, 11, 12, 22, 27, 54, 55, 56, 59, 63, 70, 77, 78, 79, 83, 86, 92, 96, 97, 98, 109, 114, 116, 117, 119, 120, 121, 131, 158, 161, 164, 168, 171
seedlings, 92
Senate, 32, 44, 92
sensitivity, 19, 118, 141, 148
September 11, 40, 177
services, 6, 8, 9, 10, 12, 15, 19, 21, 35, 48, 51, 61, 65, 73, 108, 116, 118, 150, 195
shoots, 168
showing, 14, 19, 86
shrubs, 194
Sierra Club, 80, 99
signals, 199
simulation, 148
skin, 182
social behavior, 6, 50
social interactions, 147
social order, 6, 50
solution, 128, 135, 153, 182, 199
South America, 6, 21
South Dakota, 14, 16, 31, 35
sowing, 117
soybeans, 8, 40, 56, 85, 96
specialists, 183
specialization, 87
specialty crop, 2, 5, 8, 31, 32
species richness, 118
spiders, 85
spillover effects, 22, 25
spore, 23
Spring, 98, 103, 104
stability, 117, 118
stakeholders, 29, 51, 62, 67, 71, 193, 195, 197, 199
state(s), 2, 9, 10, 12, 13, 14, 16, 19, 28, 29, 31, 35, 38, 41, 48, 51, 61, 64, 67, 70, 73, 74, 75, 75, 76, 77, 78, 80, 88, 97, 118, 194, 195, 199
state authorities, 80
state legislatures, 35, 76
statistics, 12, 13, 18, 41
statutory authority, 49
sterile, 6, 50
stock, 190, 191, 194
storage, 55, 133, 139, 140, 174
stress, 4, 16, 22, 23, 116, 131, 134, 135, 136, 147, 153, 196, 199
stress factors, 23
stressors, 4, 22, 23, 48, 58, 87, 90, 112, 140, 184, 197
structure, 108
sucrose, 89, 134
sugar beet, 56, 97
sugarbeets, 40
Sun, 98
suppliers, 86
suppression, 79
Supreme Court, 99
surface area, 174
surfactant(s), 92
surrogates, 117, 118, 160, 165
surveillance, 34

survival, 89, 112, 116, 117, 118, 130, 136, 142, 143, 146, 147, 148, 152, 162, 163, 175, 183
survival rate, 89
susceptibility, 22, 30, 59, 89, 197
suspensions, 83
sustainability, 43, 182
Sustainable Development, 98
sweat, 7, 8
symptoms, 90
syndrome, 23, 184
synergistic effect, 70, 87
synthesis, 146

T

talc, 86, 89, 117
target, 58, 64, 74, 84, 123, 127, 137, 160, 196, 197, 198
Task Force, 26, 66, 187, 189
taxa, 18, 108, 110, 116, 117, 141, 147, 152
taxonomy, 18
technical assistance, 35
techniques, 14, 64, 132, 184, 190, 191
technology, 148, 194
temperature, 21, 22, 44
test data, 155
testing, 87, 123, 124, 125, 126, 127, 128, 130, 135, 136, 144, 146, 148, 155, 183, 198
thermoregulation, 29
threats, 20, 23, 33, 34, 48
toxic effect, 90
toxicity, 53, 54, 55, 58, 59, 61, 109, 111, 112, 118, 123, 124, 125, 126, 127, 128, 129, 133, 137, 138, 144, 145, 146, 147, 148, 152, 164, 165, 168, 178, 194, 198
toxicology, 179
toxin, 76
trade, 81
traits, 19, 29, 184
transformation, 126
transformation product, 126
translocation, 150, 157, 158
transpiration, 161
transport, 74, 150, 160, 168, 169, 171
transportation, 4, 22, 38
treatment, 26, 54, 56, 59, 62, 63, 70, 71, 73, 74, 77, 83, 114, 120, 128, 130, 131, 135, 147, 153, 171, 173, 174, 175, 176, 196
triggers, 123

U

U.S. agricultural production, vii, 1, 3, 8
U.S. Department of Agriculture (USDA), 2, 3, 47, 49, 196
U.S. Geological Survey, 43
unacceptable risk, 85
unforeseen circumstances, 199
uniform, 76, 118, 147, 174
United, 1, 3, 6, 7, 8, 9, 10, 12, 14, 16, 17, 18, 19, 20, 23, 26, 27, 28, 34, 39, 42, 47, 49, 50, 53, 54, 56, 78, 79, 87, 90, 91, 92, 99, 102, 177, 179, 181, 182, 184, 188
United Kingdom (UK), 19, 20, 78, 98, 106
United Nations (UN), 3, 8, 39, 40, 91
United States, 1, 3, 6, 7, 8, 9, 10, 12, 14, 16, 17, 18, 23, 26, 27, 28, 34, 39, 42, 47, 49, 50, 53, 54, 56, 79, 87, 90, 92, 99, 102, 177, 179, 181, 182, 184, 188
universities, 16, 29, 30, 68, 90, 93, 184
updating, 75
urban, 197
USGS, 40

V

valuation, 39
vapor, 117
variables, 90
varieties, 64, 151, 188
vegetables, 8, 181
vegetation, 25, 27, 66, 129, 171, 174
vertebrates, 23, 88
viral diseases, 23, 84
viral infection, 89
viral pathogens, 25
viruses, 48, 58, 81, 82, 182, 184, 196

volatilization, 61
vulnerability, 39

W

walking, 89
Washington, 14, 16, 44, 75, 76, 93, 94, 104, 177, 178, 179, 190
water, 22, 59, 62, 63, 64, 129, 149, 161, 170, 172
water quality, 64
websites, 80
well-being, 5, 32, 49, 82
wellness, 48, 49, 51
White House, 43, 94
White Paper, 107, 108, 116, 118, 119, 120, 122, 124, 125, 130, 133, 135, 136, 137, 141, 144, 145, 148, 152, 177
wholesale, 194, 195
wild bees, vii, viii, 1, 6, 8, 9, 18, 20, 33, 49, 50, 82
wildfire, 36
wildlife, 2, 38, 75
wind speed, 117
Wisconsin, 14, 16, 31, 35
wood, 50
workers, 119, 122, 165, 197
worldwide, 6, 7, 8, 49, 50, 182

X

xylem, 120, 150, 171, 177

Y

Yale University, 45
yield, 12, 53, 70, 89, 160